Practical Design of Steel Structures

Practical Design of Steel Structures

Based on Eurocode 3 (with case studies):
A multibay melting shop and finishing mill building

Karuna Moy Ghosh

Published by
Whittles Publishing,
Dunbeath,
Caithness KW6 6EY,
Scotland, UK

www.whittlespublishing.com

Distributed in North America by
CRC Press LLC,
Taylor and Francis Group,
6000 Broken Sound Parkway NW, Suite 300,
Boca Raton, FL 33487, USA

ISBN 978-1904445-92-0
USA ISBN 978-1-4398-3571-5

Front cover photograph by kind permission of Cliff Chamberlain, www.3c-communications.co.uk

Printed and bound in England www.printondemand-worldwide.com

Contents

Preface

There are numerous books available discussing the analysis and design of steel structures. These texts consider isolated parts of a structure, with the emphasis primarily on theory and little focus on practical design and the considerations and challenges that engineers face in the design office and on the construction site.

This book takes a holistic approach presenting a comprehensive description and explanation of the analysis and design process for any structure and its component structural elements, from the initial design concept through to final construction.

The text has been written with the structural calculations presented in a simple and lucid way. Taking a step-by-step approach, the book discusses design philosophy, functional aspects of structure, selection of construction material and accompanying methods of construction, with reference to the relevant clauses of codes of practice. Included are design sketches, tables and references.

For illustrative purposes, a specific structure (accompanied by detailed worked examples) has been selected and is outlined below:

A multibay melting shop and finish mill building—The multibay melting shop and finish mill building is a complex structure that houses several heavy duty overhead electric travelling cranes. These have high vertical dynamic impact (40%) and 10% horizontal transverse crane surge on the crane girders, subsequently transferring impact on the supporting members. The structural members have been analysed and designed to resist the above dynamic impact forces.

This book describes the practical aspects of analysis and design based on the latest steel structure design codes of practice **Eurocode 3:** Part 1-1 and Part 1-8: *Design of steel structures for buildings and Design of joints.* Included is the comparative analysis of results for model design of a beam and column applying Euorocode 3 and BS 5950. 2000. The following relevant Eurocodes applicable to the analysis have also been included: **Eurocode 0:** *Basis of structural design* (BS EN 1990:2002) and **Eurocode 1:** *Densities, self weights, imposed loads, snow loads, wind loads, and cranes and machinery.*

This book will be invaluable as a practical design guide and reference text book for final year university students, newly qualified university graduates, practising engineers, consulting engineers working in the design office and at the construction site, and for those appearing for professional examinations.

Author's Note
To facilitate ease of calculation and compliance with the code, equation numbers provided in the text are those used in the code.

CHAPTER 1

General Principles and Practices

This book considers theory and its application in the context of the analysis and design of structures, addressing in particular the behaviour of the structural elements of a multibay steel-framed industrial building under the actions of heavy moving loads due to electric overhead travelling (EOT) cranes and wind forces. The analysis and design of the structural members are done in compliance with the Eurocodes, with case studies included.

Before discussing analysis and design, we must first plan the structural arrangement in relation to the requirements of the layout of machines and equipment. We must then consider the selection of construction materials, taking account of availability and cost within the scheduled construction programme and budget. We must also examine the buildability of the structure with regard to space restrictions, the method of construction, the location, ground conditions and seismic information about the site.

The above points will be discussed in detail in Section 1.2.

1.1 Brief description of the structure

1.1.1 Structural arrangement

The building complex comprises a multibay melting shop and finishing mill building. The melting shop consists of a melting bay, a hopper storage bay, an intermediate bay and a casting bay. The finishing shop consists of a rolling shop bay, a finishing mill bay and a motor and power room bay. The finishing shop is located adjacent to the melting shop, as shown in plan in Figs 1.1 and 1.2. The two buildings are separated by an expansion joint. The spacings of the stanchions in the melting bay, storage hopper bay, intermediate bay and casting bay are 28.5, 12, 27 and 30 m, respectively. In the finishing shop, all columns are spaced at 30 m centres. In the melting shop, the height of the building to the eaves level is 35.5 m except for the hopper bay, where the eaves level from the floor level is 45.5 m. In the finishing shop, the height of the building to the eaves level is 22.5 m from the ground floor. Both buildings house overhead electric cranes, as shown in section in Figs 1.3 and 1.4.

1.1.2 Overhead electric travelling cranes

Overhead electric cranes, of capacities ranging from 290 t (2900 kN) to 80 t (800 kN), run through the melting, intermediate and casting bays. The finishing shop carries cranes of capacities ranging from 80 t (800 kN) to 40 tons (400 kN). The storage hopper bay consists of hoppers storing heavy briquette iron, coke and limestone to supply to the melting furnace during operation (see Figs 1.3 and 1.4).

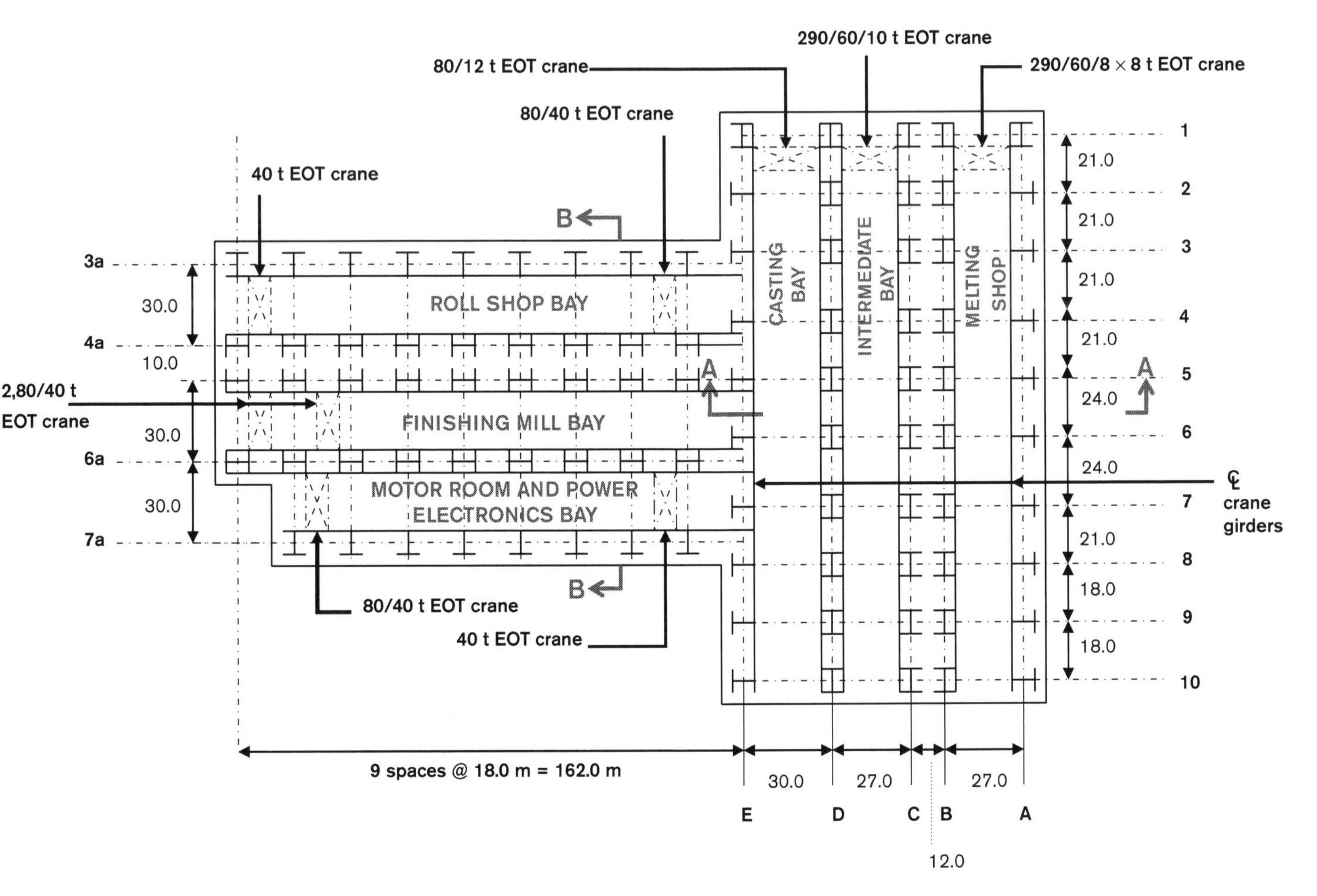

Fig. 1.1. Plan at crane girder level (showing layout of columns and cranes)

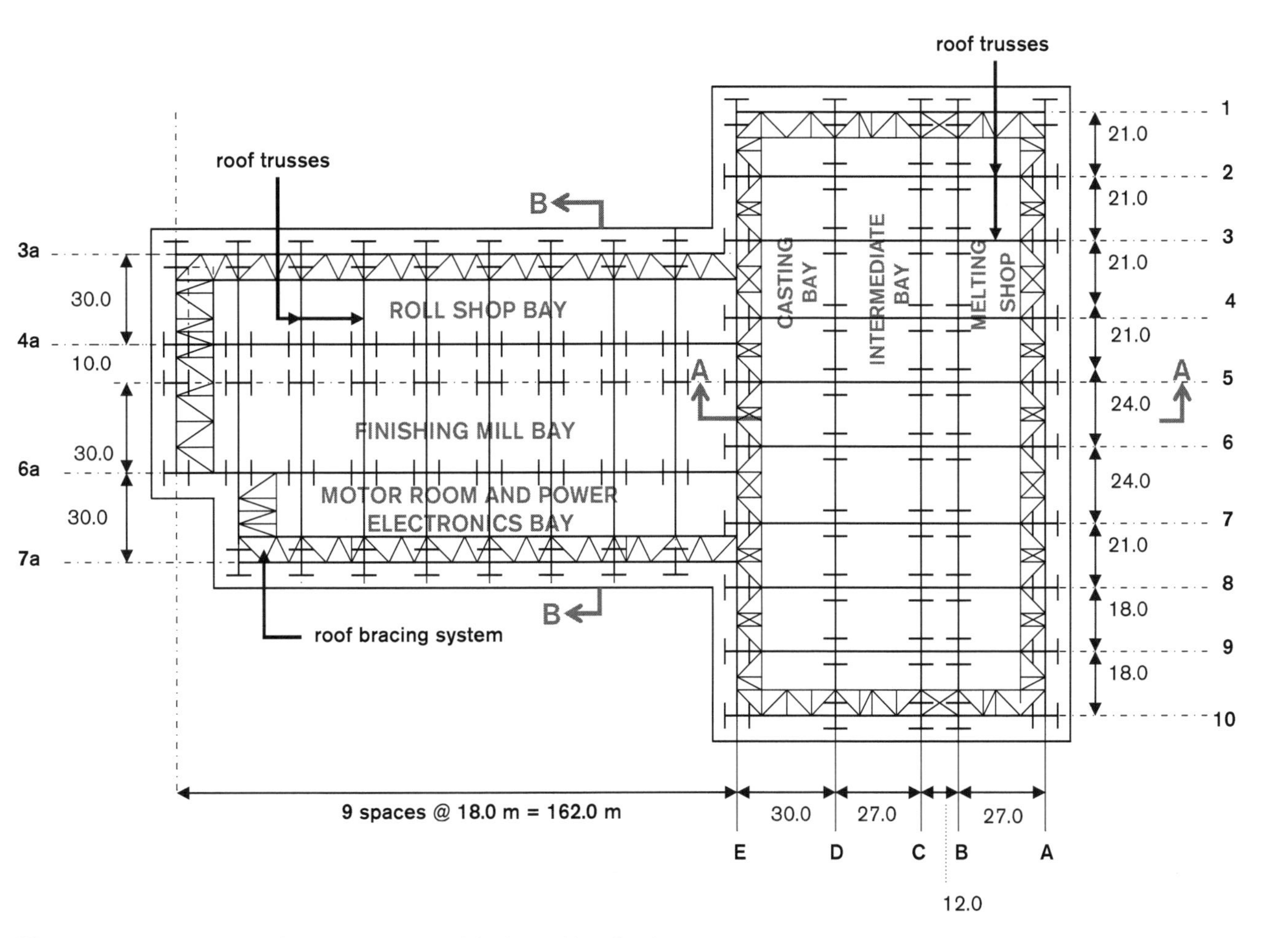

Fig. 1.2. Roof plan (showing roof trusses and horizontal bracings)

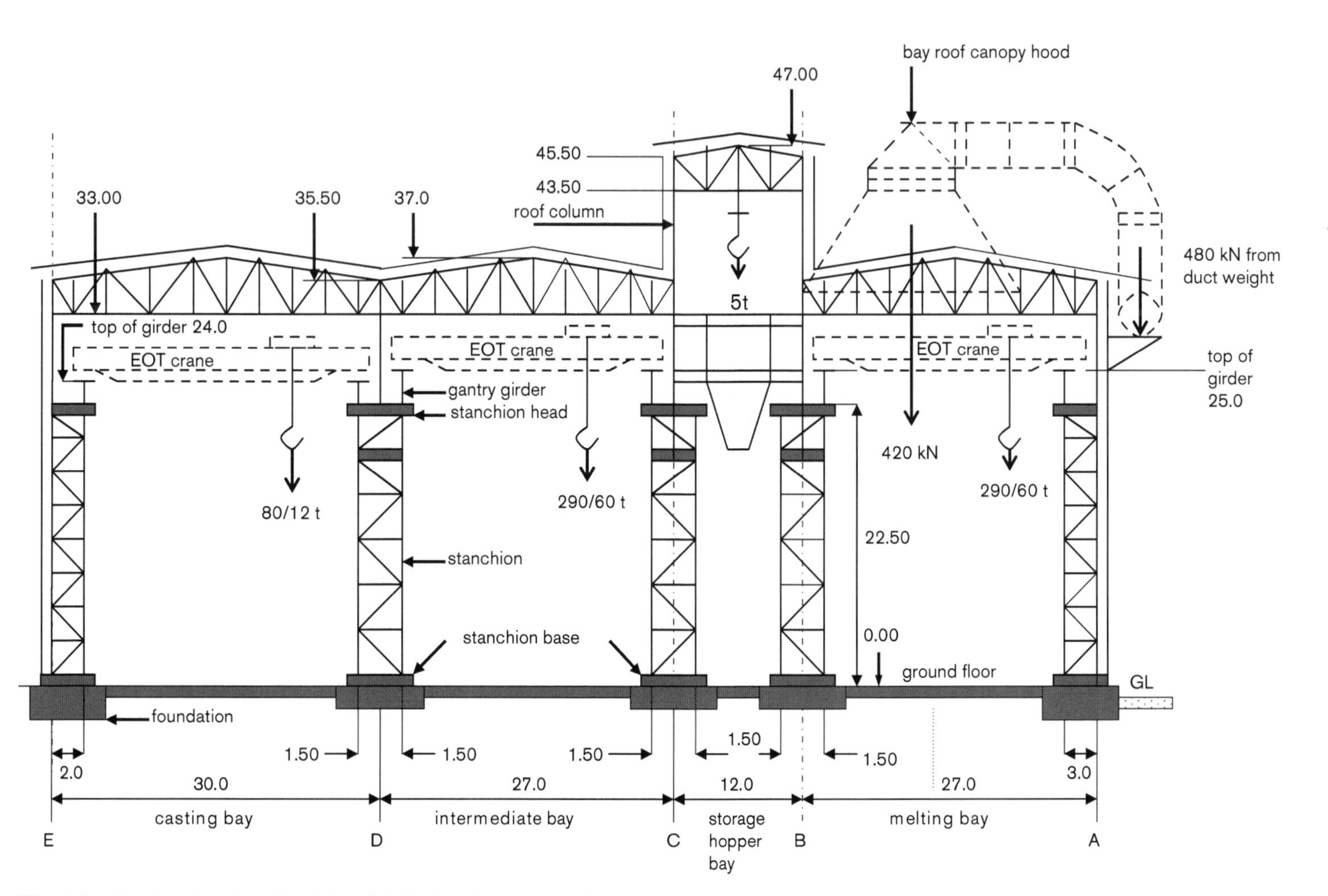

Fig. 1.3. Section A.A (see Figs 1.1 and 1.2), showing structural arrangement

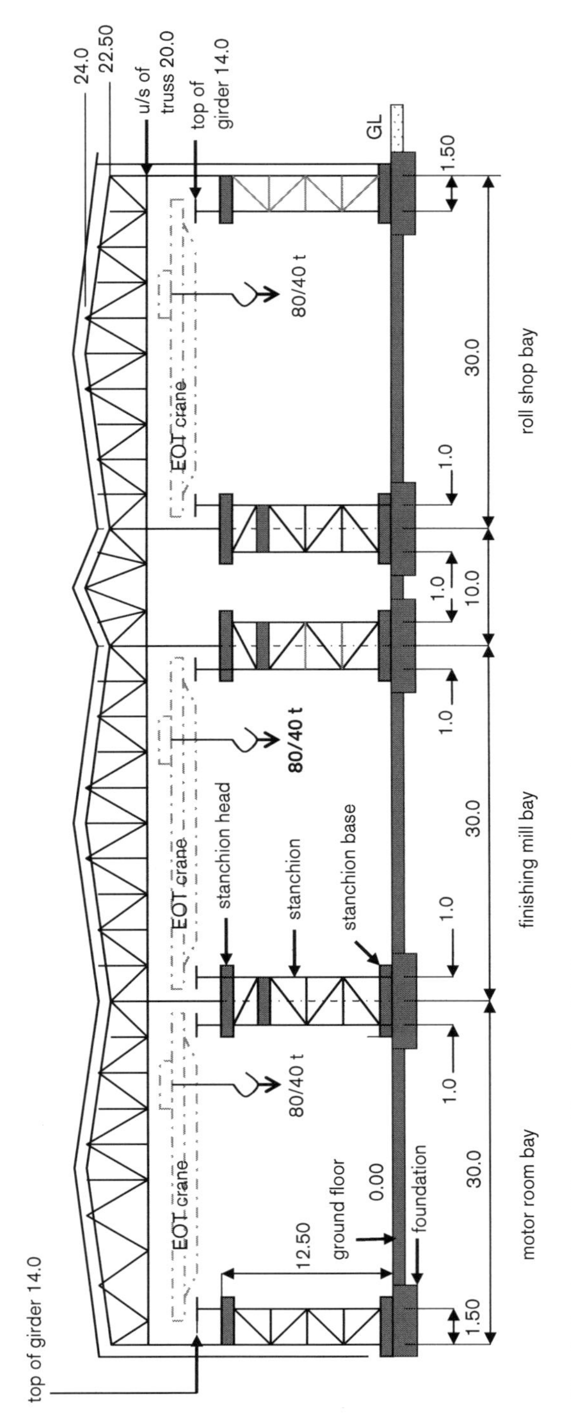

Fig. 1.4. Section B.B (see Figs 1.1 and 1.2), showing structural arrangement

1.1.3 Gantry girders

The top level of the gantry crane girders in the finishing shop is 14 m from the operating-floor level, and in the melting shop the top of the gantry crane girders is 25 m from ground level in the melting and intermediate bays and 24 m from ground level in the casting bay.

1.1.4 Fabrication of structural members

The crane girders in the melting and intermediate bays are of built-up welded-plate girder construction, and in the rest of the bays normal universal beams are used. The stanchions and roofs are of fabricated trussed-type construction and shop fabricated. Horizontal and vertical bracing systems are provided along the horizontal and vertical planes of the roof trusses and stanchions to resist wind and crane surges, respectively (see Fig. 1.5).

1.2 Design philosophy and practice

Before we proceed with the actual analysis and design of the structure, we need to consider the following aspects in order to reach a satisfactory solution to the problem:

- the functional aspects of the structure;
- alternative structural arrangements and choices of spacing of the columns and frames or trusses;
- the structural system and type;
- the buildability of the structure;
- the choice of an open or covered structure;
- the selection of the construction materials;
- the choice of shop or site connection of the component steel structures;
- the sequence and method of erection of the steel structures;
- the location, ground conditions and seismic information;
- the environmental impact of the structure;
- the design concept.

The above aspects must satisfy the requirements of Eurocode 3 and other relevant Eurocodes.

1.2.1 Functional aspects of the building

This building plays a vital role in the production of finished steel products. The melting and finishing shop, built adjacent to each other, form an important heavy industrial building complex with heavy cranes running throughout the building when it is operational. Within the building, a conveyor system supplies materials such as heavy briquette iron, coke and limestone. These materials are stored temporarily in bins hanging in the storage hopper bay. During operation, these raw materials are fed into the furnace through a conveyor system. The molten metal is then transferred to the casting bay by cranes. From here the product is transported to the finishing bay to be used in the production of continuous plate, which is rolled in the roller bay. The final product is transferred to the storage building.

The whole operation is automatically controlled from a control room adjacent to the finishing bay. Power is supplied from a generator situated in the motor and power house bay.

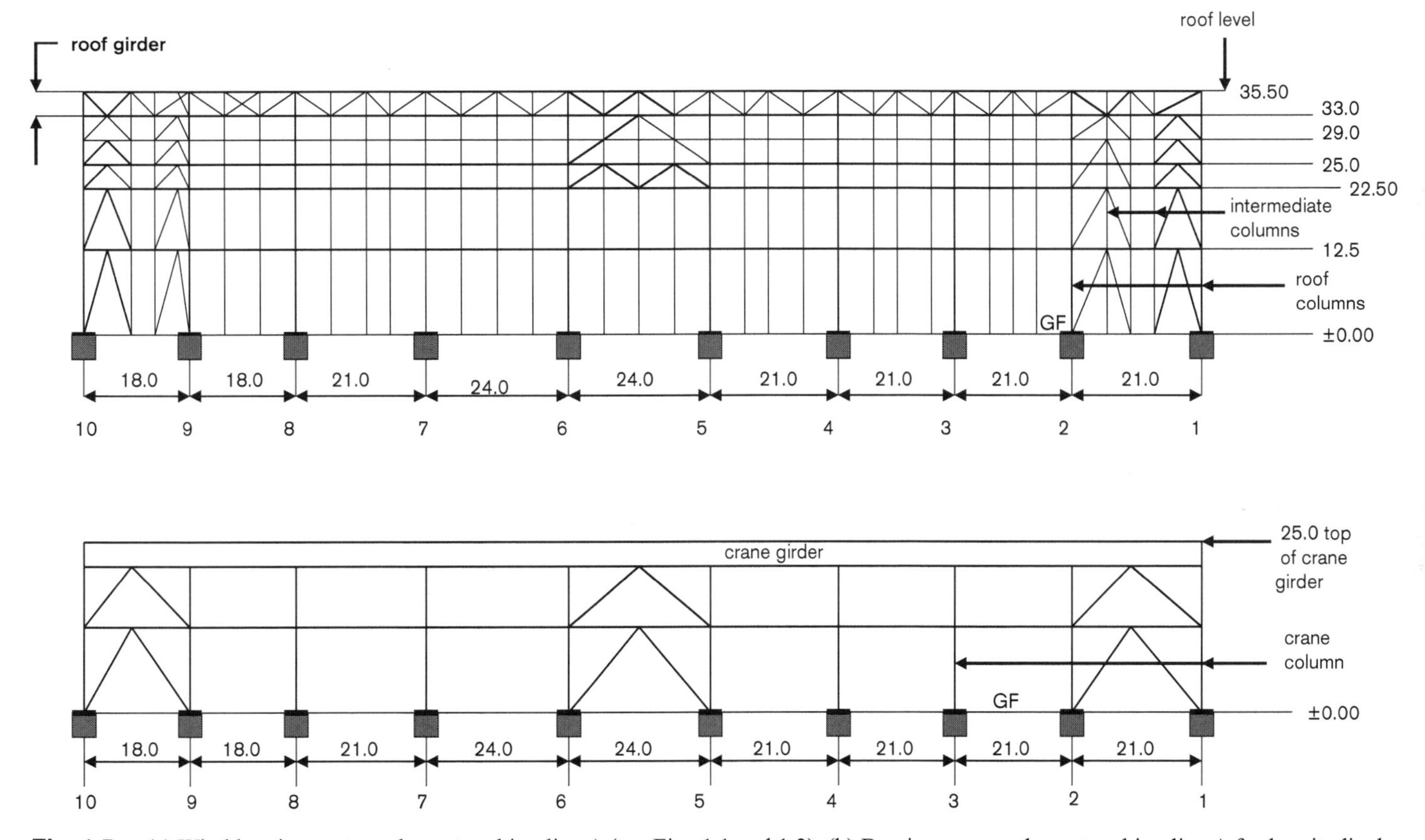

Fig. 1.5. (a) Wind bracing system along stanchion line A (see Figs 1.1 and 1.2). (b) Bracing system along stanchion line A for longitudinal active force from crane

1.2.2 Alternative structural arrangements and spacings of frames

The choice of the structural arrangement between alternative possible arrangements and the choice of the spacing of the frames are dependent mainly on the mechanical layout of the machines and equipment. They are also dependent on finding the most economical methods for mechanical handling, operations and the movement of raw materials and final products. The storage and transfer of materials are generally unidirectional.

As a result, the frames of the building structure should be oriented at right angles to the direction of movement of the conveyor system and cranes, enabling the storage and transportation of the raw materials fed into the furnace, and transfer of the final shop products to storage.

1.2.3 Structural system and type

Structural systems may be classified as follows:

- simple system of construction;
- semi-continuous (semi-rigid) construction;
- continuous (rigid) construction.

In the *simple system* of construction, the members behave as hinge connections at joints so as not to develop any moment at the joint or to transfer moments to the adjacent members. The structure should be braced in both directions to provide stability against sway. This type of system is easy to analyse and is generally used in simple types of construction where the structural arrangement is not restricted from the point of view of availability of space and architectural appearance, and also is not subjected to complicated loadings.

In the *semi-continuous* (or semi-rigid) system, the joints are assumed to have some strength and stiffness so that they can transfer moments between members. In practice, it is difficult to quantify the moment transfer between the members. This system is used in special circumstances when we are able to correctly quantify the stiffness of the joints.

In the *continuous* (rigid) system of construction, the joints between members are considered to have full rotational stiffness and rigidity so that they can transfer moments and forces between members, and to be capable of resisting moments and forces. This type of construction is represented by portal frame structures. The frames are designed and fabricated with particular attention to the joints between the members. In industrial buildings, rigid portal frame structures are frequently used where space restrictions are the main consideration.

In the above types of construction, the members, either solid universal beams or members composed of bracings, are chosen to comply with requirements on the span/depth ratio to limit the deflection of the members.

In our case, frames consisting of stanchions and roof trusses are arranged at certain spacings along the length of the building to suit the machine and equipment layout. The heavy overhead electric cranes, supported on gantry girders, travel along the length of the bays to feed the furnaces and machines and for maintenance of the furnaces and machines. The stanchion and truss connections are assumed to be hinged at the top of the stanchions, and the bases of the stanchions are assumed to be fixed.

1.2.4 Buildability

The structural layout and design of the building must be chosen based on the buildability of the structure from the point of view of fabrication, erection, maintenance and services with minimum interference with existing adjacent structures and with regard to the operation of the equipment inside the building.

1.2.5 Choice of open or covered structure

This item depends on environmental requirements and susceptibility to weather. In many industrial projects, the machines and equipment do not need any enclosure to guard against weather conditions during operation or maintenance. In our case, however, the building complex houses heavy machines and equipment for the melting and finishing of steel products. Therefore, the building needs roof and side enclosures during operation and maintenance to guard against all weather conditions.

1.2.6 Selection of construction materials

In general, two main types of construction material are in use in most industrial and service sector projects, namely *steel* and *concrete*. The selection of either steel or concrete as a building material depends on the following factors:

- *Easy availability*. The procurement of steel is easier in industrialized countries than in developing countries, whereas in developing countries concrete ingredients are easier to procure than steel.
- *Construction costs*. In developing countries, owing to the limited availability of steel, the difficulties of procurement within the scheduled project timeframe and of finding a suitable fabricator often cause delays, giving rise to escalation in the project construction costs. Completion of a structure within the scheduled timeframe provides benefits with respect to production costs for an industrial plant.
- *Financial aspects of the structure*. Consideration of the utilization of the building during the operational sequence is vital in the case of a process plant.
- *Sustainability*. Compared with other construction materials, steel possesses characteristics that lead it to being more sustainable as a construction material.

The structural complex under consideration is to be built in a developing country. The procurement of steel within the timeframe of a project is sometimes difficult; however, ensuring that the appropriate construction material is used is paramount when the functions of the building are considered. Moreover, selecting a material that supports speedy construction avoids delays to the project, which consequently impacts on production and operational costs; the construction of the building lies on the critical path and is considered to be in the sequence of the operational process. Also, the structure carries very heavy overhead cranes with appreciable degrees of mechanical vibration during operation, thus generating some fatigue stress in the material.

Considering the above points, we conclude that we should choose steel as the construction material for the above building.

1.2.7 Choice of shop or site connection of steel structures in fabrication and erection

The choice between connection in the shop and on site depends on the following factors.

Fabrication facility and capacity of fabrication shop. Generally, the structures for small and medium-size buildings are of simple braced and hinged or semi-rigid types of construction, consisting of roof trusses, universal beams or trussed girders, and universal columns with horizontal and vertical bracings to resist sway forces. These types of structures are shop fabricated with riveted and bolted connections or, sometimes, welded connections.

In industrial projects, the plant layout dictates the structural layout and the arrangement of structural members. In many cases, braced types of structure may have to be modified, with curtailment of bracings to accommodate the layout of machines and equipment. So, modern steel structures for industrial projects are designed as continuous portal frames to provide adequate clearance for the plant. This type of framing system is mainly of welded steel construction with a neat architectural appearance, and reduces the tonnage of steel required and the cost of fabrication.

Many fabrication shops do have not proper welding and testing facilities or a sufficient number of certified welders. There may also be difficulties in fabricating the whole frame as one unit to act a rigid framed construction, as it occupies a large surface area of the shop floor, which many fabrication shops cannot afford to provide. Therefore we need larger floor shop areas and adequate fabrication equipment and machines to maintain the optimum fabrication facilities and capacity of the shop.

Facilities for transport of fabricated structural components from the shop to the site. It is not always possible to transport a whole frame unit or structural component from the fabrication shop to the erection site, because of restrictions related to transport vehicle dimensions and highway regulations. So it is general practice and convenient to fabricate a frame in parts to facilitate easy transportation. In special circumstances, a fabrication shop is erected on site to meet the erection schedule.

Portal frames made up of single universal sections or composite sections of universal columns braced together by angles or channels are fabricated in three or more parts, namely:

- *Columns with bases.* Each column, its base and a small portion of the end part of the rafter are welded together. For composite sections, the bracings are welded. In the event that the length of the composite section makes the section too heavy or long to handle, it may be convenient to make the section in two parts; these are then connected on site. The bracings at the connection point are transported loose and site connected with high-strength bolts along with the main column members. The length of the part should preferably not exceed 12 m as the length of a trailer is generally limited to 12–13 m to satisfy highway regulations. In special circumstances, longer lengths may be allowed with special highway authority permission.
- *The central rafter (beam) portion.* The section at which the rafter is cut off should be at the position of minimum bending moment, and there should be provision of moment connections to the column ends at the ends of the rafter during erection to form the full portal frame. In the case where the rafter is of truss or lattice girder construction, the rafters are shop fabricated in parts of allowable length for transport to the site. They are then assembled on site to full span length, and are provided at the ends with moment connections to the columns during erection.

1.2.8 Sequence and method of erection of steel structures

The sequence and method of erection are generally dependent on the layout and arrangement of the structural components. Normally, the following sequence and method of erection are followed.

Stage 1. Before the erection of steelwork starts, the sizes and exact locations of the holding-down bolts on the foundation and base plates are checked, as often discrepancies occur that delay the erection schedule. After these checks are made, the following steps are carried out:

- Erect the columns together with base plates.
- Align the columns.
- Adjust the holding-down bolts with adjustable screws underneath the base plate to keep the required gap for grout between the underside of the base plate and the foundation.
- Use temporary bracings to hold the columns truly vertical so that they do not sway in any direction.
- When the columns or stanchions are very long, these are erected in sections and bolted together on site.

Stage 2:

- Erect the central portion of each rafter. In the case of a roof truss, the whole section may be erected in one piece.
- Connect the truss to the column ends with bolts to form the full structural frame.

Stage 3:

- Install the vertical column bracing and roof bracings after the final alignments and adjustments of the frame positions to make the whole structure stable.

Stage 4:

- Fasten all roof purlins and sheeting rails to the structure with bolted connections.

Stage 5:

- Erect crane girders where applicable.

Stage 6:

- Install cranes where applicable.

Stage 7:

- Fix the roof and side coverings.

Stage 8:

- Finally, fill the underside of the base plates with non-shrink grout after completion of erection.

The above sequence of erection is applicable to normal construction. In the case of special structures, a special sequence should be followed after the preparation of an erection program.

1.2.9 Location, ground conditions and seismic information

1.2.9.1 Location

The process of steelmaking demands a very high quantity of water for cooling the final product. A continuous supply of water is essential to keeping production running. In the planning stage, the location of the steel plant should be selected with regard to proximity to a continuous source of water. Therefore, in this project, the site was located near the estuary of a river, in a developing country as previously mentioned.

1.2.9.2 Ground conditions

In the initial stages of planning, when the layout of the structural system is considered in relation to the layout of the machines and equipment, one of the most important factors in selecting the type of foundation is to determine the ground conditions and the strength of the soil on the site. To determine the strength of the soil, subsurface ground explorations of the building project site are carried out to establish the soil properties at different depths. This is done by boring holes at marked locations and collecting borehole logs of undisturbed soil samples and by digging shallow pits for field tests. After the site explorations are complete, the borehole logs of the soil samples are taken to a soil laboratory and various tests are carried out to obtain the characteristics of various soil strata at different depths. Then the laboratory test results are recorded in tabular form for ready reference.

For example, Table 1.1 shows the geotechnical soil parameters of the ground where the foundations of the multibay melting shop and finishing mill building considered here will be constructed. Referring to Tables 1.1 and 1.2 (the latter showing the relation between the SPT N value and the allowable bearing pressure), we find that, up to a depth of 10 m below ground level, the soil has very low values of the angle of internal friction and of N and N_q, so the soil has little shearing and bearing strength. When the ground in the building area is subjected to high overburden pressure due to the piling up of raw materials, or where there are very high concentrated loads from the superstructure transmitted to the foundation level, we have to determine the strength of the soil at various depths.

Based on the above geotechnical reports from the soil investigation, we find that at shallow depths the soil has a low shearing and bearing strength. To construct an isolated

Table 1.1. Soil parameters of strata at various depths[a]

Soil type	Depth below ground level (m)	γ_b (kN/m^3)	φ' (°)	δ (°)	K_a	K_p	K_0	N	N_q
Topsoil	0	16	0	0	0	0	0	0	0
Very soft silty clay with fine sand	2.5	18	12	0	0	0	0	4	4
Medium to stiff clayey fine sand	10	20	30	22.5	0.29	5.0	1.0	1.0	10–20
Dense to very dense grey sand	35	20	35	26.3	0.22	6.2	1.0	27	35–40
Stiff silty clay with fine sand	48	20	40	30	0.18	8.0	1.0	45	60–90

[a] γ_b = density of soil, φ' = angle of internal friction, δ = angle of friction between soil and contact surface of foundation, K_a = Rankine active pressure coefficient = $(1 - \sin\varphi')/(1 + \sin\varphi')$, K_p = Rankine passive pressure coefficient = $(1 + \sin\varphi')/(1 - \sin\varphi')$, K_0 = coefficient of horizontal earth pressure at rest, N = number of blows/300 mm in standard penetration test (SPT), N_q = bearing capacity factor.

Table 1.2. Relationship between SPT N value and relative density of clayey soil

N value (blows/300 mm of penetration)	Relative density	Allowable bearing value at 4 m depth (kN/m^2)
0–4	Very soft to soft	0–30
4–8	Medium	30–100
10–20	Medium to stiff	100–300
20–40	Dense to very dense	300–500
40–60	Stiff silty clayey sand	500–650

footing foundation, we would have to build up the foundation from deep below ground level to obtain adequate soil strength to carry the required load. This type of foundation would be too costly and uneconomical. The easy solution is to adopt piled foundations (comprising a series or group of piles) to support the loads and to extend the piles deep into substrata of sufficient shearing and bearing resistance. In other cases, when the soil has adequate shearing strength at a shallow depth, it is economical to construct an isolated footing for each foundation loading.

In our case, the stanchion bases carry very high loads from the superstructure. In order to support these very heavy loads, an isolated footing foundation of the normal kind would have to be constructed at least 10 m below ground level to achieve the required shearing strength of the soil. This type of construction is costly and uneconomical. The alternative and economical way to solve the foundation problem is to use piles and take them down to a soil stratum of adequate shearing and bearing strength to support the heavy base load of the stanchions. Therefore we shall adopt a piled foundation consisting of a group of piles of sufficient depth to attain adequate shearing and bearing capacity, with a pile cap to support the bases of the stanchions, which are subjected to heavy loads and horizontal thrust.

1.2.9.3 Seismic information

Earthquakes are natural phenomena that generate dynamic ground wave motions and often cause massive devastation of buildings, bridges and dams and loss of human life. In the planning stage, very careful consideration should be given to selecting the location of structures. The pattern of distribution of earthquakes over the earth was recorded in the International Seismological Summary (US Geological Survey, 1963). Since 1964, updated information has been published at intervals in the *Bulletin of the International Seismological Centre.*

From an analysis of seismological data, however, it was found that the region under consideration is in an area of minimal or no seismic activity, and hence we shall not consider any seismic forces on the structure.

1.2.10 Environmental impact

The structure is a process plant building and, as a result, there will be considerable carbon emissions from it. Therefore every effort should be made to minimize carbon emissions in order to make the plant more environmentally friendly.

1.2.11 Design concept

This is the most fundamental aspect of the analysis and design of a structure and of its structural elements. We need to have a clear idea of how the structure will behave under various types of loadings, points of application of loadings, and sequences of loadings. Accordingly, we have to arrange the structural components that constitute the whole structural unit, bearing in mind the requirements on the mechanical layout. The structural arrangement should be made simple so that the analysis can be carried out without too much complexity; at the same time, the structure must be buildable in conformity with standard requirements.

References

US Geological Survey, 1963. *International Seismological Summary: Earthquake Catalog (1918–1963),* US Geological Survey, Reston, VA, USA.

CHAPTER 2

Structural Analysis and Design

2.1 Structural analysis

A clear idea of the behaviour of a structure and structural members subjected to various types of forces forms the basis of structural analysis. The structural components that constitute the whole structure should be built to standard requirements, in our case conforming to Eurocode 3 and other standard specifications. Sometimes, the structural system is portal framed (continuous) with rigid connections and without bracings, to provide clearance for the requirements of the layout of machines and equipment, and to create more working space. In general, portal-framed structural systems reduce the size of structural components, hence reducing the space required and minimizing cost.

2.2 Methods and procedures for analysis and design

2.2.1 Methods of analysis

The following methods of analysis are generally used in practice:

- manual analysis using fundamental theories and applications of those theories;
- manual analysis using standard formulae;
- analysis with the aid of structural-software programs;
- structural modelling.

Manual analysis using fundamental theories is the most reliable method but is time-consuming and costly, particularly when the structural system and the application of forces on the structure are complicated.

Manual analysis using standard formulae is faster, but the results should be spot-checked by applying basic theories. This method is useful for obtaining a rough estimate of costs for submitting tender documents for work, and for feasibility studies.

Analysis with the aid of structural-software programs is carried out for complicated structures with several redundant members and when time is of the essence in the final construction stage. There are many off-the-shelf programs, so it is important that the program used is reliable and well established. These types of programs should be used by experienced engineers who have a thorough knowledge of the interpretation of the output results. The output results should be verified occasionally by using fundamental theories and standard formulae.

Structural modelling for analysis in accordance with Eurocode 3, Part 1-1, BS EN 1993-1-1: 2005 (Eurocode, 2005a). Analysis is carried out based on structural modelling with consideration of ultimate-limit-state methods. The inputs to the structural model should use basic assumptions in the calculations and should reflect the structural behaviour in the limit state with sufficient accuracy to reflect the anticipated type of behaviour of the cross-sections, members, joints and bearings. The method used for analysis should be consistent with the design assumptions.

2.2.2 Procedures for the analysis

The following steps should be followed in the analysis process:

Preparation of a suitable and workable structural model to meet the requirements of machines, equipment and the environment, as well as the loadings.

Assume the design parameters.

Assume the section and size of structural components.

Consider load cases for each type of load or force.

Carry out analysis for each characteristic load case by any suitable method.

Compare and check the results with manual computation when the analysis is done with the aid of a structural program.

2.2.3 Procedures for the design of structural members

The following steps should be followed in the design of structural members based on Eurocode 3, Part 1-1 and Eurocode 0, BS EN 1990: 2002(E) (Eurocode, 2002a):

Obtain the results from the analysis.

Prepare diagrams for the characteristic bending moments, shears and axial forces in the members.

Based on Eurocode 0, combine the results for various load combinations with the appropriate partial safety factors given in the relevant codes of practice to arrive at the maximum ultimate design values for individual members.

Based on Eurocode 3, Part 1-1, design the sections of members based on the ultimate-limit-state (ULS) method for the strength of members and also on the serviceability limit state (SLS) method for deflection in accordance with relevant codes of practice.

Worked examples of model designs of a beam and column are provided in Section 2.8.

2.3 Design data

Before we start the analysis and design of a structure, we shall discuss in general various types of loadings, the intensity of loadings and the point of application of loadings on structures, and the codes of practice to be followed to obtain assumptions about loadings with partial safety factors, for load combinations that will result in the ultimate design loadings on the structure. We shall also discuss the allowable stresses in steel subjected to various internal stresses in the structural members. In addition, we shall give a general specification regarding the span of structural members, and the form, pitch and spacing of roof trusses which govern the design of gantry girders and lifting beams.

2.3.1 Loads

2.3.1.1 Dead loads (g_k in kN/m^2), based on Eurocode 1, Part 1-1, BS EN 1991-1-1: 2002] (Eurocode, 2002b)

Corrugated galvanized steel sheeting (weight in kN/m^2)

We assume the sheeting to have two corrugations of side lap and 150 mm of end lap. For 18 BWG sheeting (1.257 mm thick), weight = 0.15 kN/m^2. For 20 BWG sheeting (0.995 mm thick), weight = 0.12 kN/m^2. For 22 BWG sheeting (0.79 mm thick), weight = 0.10 kN/m^2. Normally, 20 BWG and 22 BWG corrugated sheets are used in roof and side coverings, respectively.

Glazing (weight in kN/m^2)

6 mm plain rolled plate glass including fixings = 0.34 kN/m^2.

Roof insulation (weight in kN/m^2)

12 mm plasterboard (gypsum) = 0.11 kN/m^2.

Service loads (lighting, sprinklers etc.)

Weight = 0.10 kN/m^2.

For all other construction materials, see Eurocode 1, Part 1-1.

Snow loads (weight in kN/m^2 of horizontal roof surface), based on Eurocode 1, Part 1-3, BS EN 1991-1-3 (Eurocode, 2003)

The snow load on the roof to be considered varies depending on the following points:

- the slope of the roof;
- the horizontal wind pressure on the roof;
- the location (and orientation) of the building.

If the minimum pitch of the roof is 1/4 (minimum angle of inclination of roof = 26°34′) and a minimum horizontal wind pressure of 1.37 kN/m^2 (equivalent to 95 miles/hour, or 152 km/hour) acts on the roof surface, the snow load on the roof may be ignored, because with such a high wind velocity the snow is blown away even on the leeward surface owing to the creation of eddy currents. However, in regions where snowfalls occur, a minimum of 0.24 kN/m^2 on the horizontal projection of the roof surface should otherwise be allowed. In severe arctic conditions, the snow load may be increased to up to 1.17 kN/m^2. In cases where the roof inclination is less than 26°34′, the snow load should be considered in the design.

2.3.1.2 Imposed loads (live load q_k in kN/m^2), based on Eurocode 1, Part 1-1, BS EN 1991-1-1: 2002

On roof

With access: if the slope does not exceed 10°, uniformly distributed load (UDL) = 1.5 kN/m^2. Without access, UDL = 0.75 kN/m^2.

Inspection walkways

UDL = 1.0 kN/m^2 (minimum).

2.3.1.3 Moving wheel loads from overhead electric travelling cranes, based on Eurocode 1, Part 3, BS EN 1991-3: 2002 (Eurocode, 2006)

Dynamic vertical wheel load

As a result of a sudden drop of a full load, a slip of the sling or a sudden braking action during the travel of a fully loaded crane (where the load includes the self-weight of the crane), a dynamic effect on the wheels is generated, thereby increasing the static wheel load. This effect is defined by an impact factor that is multiplied by the static wheel load to give the dynamic wheel load. Thus

maximum dynamic vertical wheel load = static wheel load × dynamic factor (φ).

The dynamic factor varies depending on the class of duty (loading class) of the crane. Table 2.1 gives various loading classes. Groups of loads and dynamic factors to be considered as one characteristic crane action are listed in Table 2.2. The equations listed in Table 2.3 should be used to calculate the values of the dynamic factors φ_i.

The vertical dynamic factors can be evaluated as follows.

For hoisting classes HC1and HC2, for example, referring to Table 2.3, we have the dynamic factor φ_1 for vertical loads: $0.9 < \varphi_1 < 1.1$. Assume $\varphi_1 = 0.9$, the lower value for vibrational pulses. We also have

$$\varphi_2 = \varphi_{2,\,\mathrm{min}} + \beta_2 v_\mathrm{h}$$

where $\varphi_{2,\,\mathrm{min}} = 1.05$ and $\beta_2 = 0.17$ for hoisting class HC1, and $\varphi_{2,\,\mathrm{min}} = 1.1$ and $\beta_2 = 0.34$ for hoisting class HC2 (see Table 2.4). In addition,

V_b = steady hoisting speed = 1.3 m/s (assumed).

Therefore

for class HC1: dynamic factor = $\varphi_2 = 1.05 + 0.17 \times 1.3 = 1.27$;

for class HC2: dynamic factor = $\varphi_2 = 1.1 + 0.341.3 = 1.54$.

Thus, referring to Table 2.2 and assuming the group of loads 1, we have the following:

for class HC1: $\varphi = \varphi_1 \varphi_2 = 0.9 \times 1.27 = 1.14$;

Table 2.1. Recommendations for loading classes (based on Table B.1 in Eurocode 1, Part 3)[a]

Item	Type of crane	Hoisting class	S-class
1	Hand-operated cranes	HC1	S0, S1
2	Assembly cranes	HC1, HC2	S0, S1
3	Power house cranes	HC1	S1, S2
4	Storage cranes with intermittent operation	HC2	S4
5	Storage cranes and spreader bar cranes, with continuous operation	HC3, HC4	S6, S7
6	Workshop cranes	H2, H3	S3, S4
7	Overhead travelling cranes and ram cranes, |with grab and magnet operation	HC3, HC4	S6, S7
8	Casting cranes	HC2, HC3	S6, S7
9	Soaking-pit cranes	HC3, HC4	S7, S8
10	Stripper cranes, charging cranes	HC4	S8, S9
11	Forging cranes	HC4	S6, S7

[a] The bottom part of the table has been omitted as the types of crane in the bottom part are not relevant in this context.

Table 2.2. Groups of loads and dynamic factors to be considered as one characteristic crane action (based on Table 2.2 of Eurocode 1, Part 3)[a]

				Groups of loads									
				Ultimate limit state							Test load	Accidental load	
Item	Description	Symbol	Section of Eurocode 1, Part 3	1	2	3	4	5	6	7	8	9	10
1	Self-weight of crane	Q_c	2.6	φ_1	φ_1	1	φ_4	φ_4	φ_4	1	φ_1	1	1
2	Hoist load	Q_h	2.6	φ_2	φ_3		φ_4	φ_4	φ_4	η		1	1
3	Acceleration of crane bridge	H_L, H_T	2.7	φ_5	φ_5	φ_5	φ_5	–	–	–	φ_5	–	–
4	Skewing of crane bridge	H_s	2.7	–	–	–	–	1	–	–	–	–	–
5	Acceleration or braking of crab or hoist block	H_{T3}	2.7	–	–	–	–	–	1	–	–	–	–
6	In-service wind	F_w	Annex A	1	1	1	1	1	–	–	1	–	–
7	Test load	Q_T	2.10	–	–	–	–	–	–	–	φ_6	–	–
8	Buffer force	H_B	2.11	–	–	–	–	–	–	–	–	φ_7	–
9	Tilting force	H_{TA}	2.11	–	–	–	–	–	–	–	–	–	1

[a] η is the proportion of the hoist load that remains when the payload is removed; it is not included in the self-weight of the crane.

Table 2.3. Dynamic factors φ_i for vertical loads (based on Table 2.4 of Eurocode 1, Part 3)

Dynamic factor φ_i	Value of dynamic factor
φ_1	$0.9 < \varphi_1 < 1.1$ The two values 1.1 and 0.9 reflect the upper and lower values of vibrational pulses
φ_2	$\varphi_2 = \varphi_{2,\min} + \beta_2 v_h$, where v_h = steady hoisting speed in m/s For $\varphi_{2,\min}$ and β_2, see Table 2.4
φ_3	$\varphi_3 = 1 - \Delta m\,(1 + \beta_3)/m$, where Δm = released or dropped part of the hoisting mass, m = total hoisting mass, $\beta_3 = 0.5$ for cranes equipped with grabs or similar slow-release devices, and $\beta_3 = 1.0$ for cranes equipped with magnets or similar rapid-release devices.
φ_4	$\varphi_4 = 1.0$ provided that the tolerances for rail tracks as specified in EN 1993-6 are observed

Table 2.4. Values of β_2 and $\varphi_{2,\min}$ (based on Table 2.5 of Eurocode 1, Part 3)

Hoisting class of appliance	β_2	$\varphi_{2,\min}$
HC1	0.17	1.05
HC2	0.34	1.10
HC3	0.51	1.15
HC4	0.68	1.2

for class HC2: $\varphi = 0.9 \times 1.54 = 1.38$.

The following vertical dynamic factors may be used as guidance for various hoisting classes:

- *For hoisting class HC1*, light-duty hand-operated cranes, assembly cranes, power house cranes and intermittently used storage cranes: dynamic factor $\varphi = 1.1$ (minimum) to 1.25.
- *For hoisting class HC2*, medium-duty cranes (normally in factories, workshops and warehouses, and for casting and in scrapyards with continuous operation): dynamic factor $\varphi = 1.25$ to 1.4.
- *For hoisting class HC3*, heavy-duty cranes (in foundries and for intermittent grab and magnet work, forging, charging etc.): dynamic factor $\varphi = 1.4$ (minimum).

Generally, the crane manufacturer will provide the dynamic factor along with the crane wheel loads when details of the duty (class), the span of the crane and the lifting capacity are given to the manufacturer. In our case, the vertical dynamic factor (φ) provided by the crane manufacturer is 1.4.

Transverse horizontal force (surge) on a crane girder during travelling of crane

This transverse horizontal surge is generated owing to the following factors:

- Thrust from sudden application of the brakes of the crab motor, causing abrupt stoppage of the crab and load when traversing the crab girders. This thrust is resisted by the frictional force developed between the crab wheels and crab girders, is then transferred to the crosshead girders of the crane, and is finally transferred as point loads through the main wheels of the crane into the top flange of the crane girders.
- A crane often drags weights across the shop floor. If the weight is very heavy, this pulling action induces a transverse horizontal component of force (a point load) on the crane girders through the crane wheels.

The transverse horizontal force generated by either of the above causes or by a combination of both of them is transferred to the crane girders through the double-flanged crane wheels on the end carriages, and cranes are designed to avoid the possibility of derailment.

It is quite difficult to determine the value of this force quantitatively, as there are unknown factors besides the above facts. American specifications stipulate that the horizontal transverse force on each gantry girder is equal to 10% of the load lifted. The British code of practice BS 2573-1: 1983 (British Standards Institution, 1983) specifies the following:

value of total transverse horizontal force = $1/10 \times$ weight of (lift load + crab).

Eurocode 1, Part 3 stipulates the same value. Therefore

value of total transverse horizontal force = $1/10 \times$ weight of (lift load + crab).

This force should be shared equally between the two gantry girders.

Longitudinal horizontal force

During the travelling of the crane, the sudden application of brakes induces frictional resistance to the sliding of the locked wheels upon a rail fixed to the gantry girder. This frictional resistance, in turn, generates a horizontal force along the length of the gantry girder, and finally transfers to the columns that support the gantry girder. Assume that the coefficient of friction μ for steel sliding on steel is 0.2. Consider the maximum vertical wheel load on the gantry girder, which occurs when the load lifted is at the nearest allowable position to the gantry girder. So,

maximum wheel load on the nearest gantry girder = maximum reaction from crane (load lifted + half the dead weight of crane) = $W = R$.

For example, if the load lifted is W_1, the self-weight of the crane is W_2, the distance of the load lifted from the nearest gantry girder is l and the crane span (centre to centre of cross-head) is L, then

maximum on-wheel load = $W_1(L - l)/L + W_2/2 = W = R$.

Therefore

longitudinal horizontal force developed = $R\mu = 0.2R$.

The American code of practice specifies that the longitudinal force is equal to 10% of the maximum wheel load. The British code of practice BS 2573 specifies that the longitudinal force is equal to 5% of the maximum wheel load, assumed to be acting on one gantry girder nearest to the load lifted. Eurocode 1 stipulates that the longitudinal force applied to the gantry girder should be calculated as follows (the equation numbers given in this chapter refer to Eurocode 1, Part 3):

$$H_{L,i} = \varphi_5 K_i / n_r \tag{2.2}$$

where

n_r = number of gantry girders = 2,
K = driving force (the value should be provided by the crane supplier),
φ_5 = dynamic factor (see Table 2.5),
i = integer to identify the gantry girder ($i = 1, 2$).

We shall follow the crane manufacturer's instructions here, which specify the longitudinal force applied to the gantry girder. Therefore we adopt H_{Li} = 5% of the maximum wheel load, assumed to be acting on one gantry girder nearest to the load lifted.

Table 2.5. Dynamic factor φ_5 (based on Table 2.6 of Eurocode 1, Part 3)

Value of the dynamic factor φ_5	Specific use
$\varphi_5 = 1$	For centrifugal forces
$1.0 \leq \varphi_5 \leq 1.5$	For systems where forces change smoothly
$1.5 \leq \varphi_5 \leq 2.0$	For cases where sudden changes can occur
$\varphi_5 = 3.0$	For drives with considerable backlash

2.3.1.4 Wind pressure on buildings, based on Eurocode 1, Part 1-4 (Eurocode, 2005b)

Dynamic wind pressure (peak velocity pressure)

The dynamic horizontal wind pressure on a vertical surface is given by the equation

$$q_s = 0.613 v_e^2$$

where q_s is the dynamic wind pressure in N/m^2 and v_e is the effective wind speed in m/s. This equation is not mentioned in Eurocode 1.

In Eurocode 1, equations are given for the *peak velocity pressure* for the action of wind on structures. Therefore we shall follow Eurocode 1, Part 1-4 to arrive at the peak velocity pressure.

Effective wind velocity

To arrive at the effective wind velocity, we have to start from the basic wind velocity.

Fundamental value of the basic wind velocity ($v_{b,0}$)

The fundamental value of the basic wind velocity $v_{b,0}$ is the characteristic 10 minute mean wind velocity irrespective of wind direction and time of the year, at 10 m above ground level in open-country terrain with low vegetation such as grass, and with isolated obstacles with separations of at least 20 obstacle heights.

Basic wind velocity

The basic wind velocity is calculated from the following equation:

$$v_b = v_{b,0} c_{dr} c_{season} \qquad (4.1)$$

where

c_{dr} = direction factor, recommended value 1.0,
c_{season} = season factor, recommended value 1.0.

Therefore

$$v_b = v_{b,0}$$

Assume that the basic wind velocity v_b (obtained from meteorological data) is 24 m/s.

Mean wind velocity

The mean wind velocity at a height z above the terrain depends on the roughness and orography of the terrain and on the basic wind velocity v_b, and may be calculated from the following equation:

$$V_m(z) = v_b c_r(z) c_o(z) \qquad (4.3)$$

where $c_r(z)$ is the roughness factor of the ground roughness of the terrain upwind of the structure in the wind direction considered. The roughness factor at a height z may be calculated from the following equation:

$$C_r(z) = k_r \ln(z/z_0) \qquad \text{for} \quad z_{min} \leq z \leq z_{max} \qquad (4.4)$$

or

Table 2.6. Terrain categories and terrain parameters (based on Table 4.1 of Eurocode 1, Part 1-4)

No.	Terrain category	z_0 (m)	z_{min} (m)
0	Sea or coastal area exposed to the open sea	0.003	1
I	Lake or flat, horizontal area with negligible vegetation and without obstacles	0.01	1
II	Area with low vegetation such as grass and isolated obstacles (trees, buildings) with a separation of at least 20 obstacle heights	0.05	2
III	Area with regular cover of vegetation or buildings or with isolated obstacles with separations of maximum 20 obstacle heights (such as villages, suburban terrain or permanent forest)	0.3	5
IV	Area in which at least 15% of the surface is covered with buildings and their average height exceeds 15 m	1.0	10

$$c_r(z) = c_r(z_{min}) \quad \text{for} \quad z \leq z_{min}$$

where $c_o(z)$ is the orography factor; z is the height of the structure above ground level, equal to 47 m (see Figs 1.1–1.5); z_0 is the roughness length; k_r is a terrain factor depending on the roughness length z_0, calculated using the equation

$$k_r = 0.19(z_0/z_{0,\,\text{II}})^{0.07} \tag{4.5}$$

where $z_{0,\,\text{II}} = 0.05$ m (terrain category II; see Table 2.6); z_{min} is the minimum height, defined in Table 2.6; and z_{max} is taken as 200 m.

Assuming terrain category II, $z_{min} = 2$ m and $z_{max} = 200$ m, we obtain $z_0 = 0.05$. So,

$$k_r = 0.19[0.05/0.05]^{0.07} = 0.19$$

Also,

$$z = \text{building height} = 47 \text{ m}$$

Therefore

$$c_r(z) = k_r \ln(z/z_0) = 0.19 \ln(47/0.05) = 0.19 \times 6.85 = 1.3$$

Terrain orography

Where the orography (e.g. hills or cliffs) increases the wind velocity by more than 5%, the effects of this should be taken into account using the orography factor $c_o(z)$. However, the effect of orography may be neglected when the average slope of the upwind terrain is less than 3°. The upwind terrain may be considered to extend to a distance of up to 10 times the height of an isolated orographic feature. In our case, the average slope of the upwind terrain is assumed to be less than 3°. So,

$$c_o(z) = 1.0$$

Therefore

$$\text{mean wind velocity} = v_m(z) = c_r(z)c_o(z)v_b = 1.3 \times 1.0 \times v_b = 1.3v_b$$

N.B. If the construction area is located in a coastal zone and exposed to the open sea, the terrain should be classified in category 0. Then, from Table 2.6, $z_0 = 0.003$. The value of $c_r(z) = 0.19 \ln(47/0.003) = 1.84$ is greater than that for category II, and the mean wind velocity is given by

$$v_m(z) = 1.84 \times 1.0 \times v_b = 1.84v_b > 1.3v_b$$

Peak velocity pressure

The peak velocity pressure $q_p(z)$ at a height z is given by the following equation:

$$q_p(z) = c_e(z)q_b \tag{4.8}$$

where

$$c_e(z) = \text{exposure factor} = q_p(z)/q_b \tag{4.9}$$

$$q_b = \text{basic velocity pressure} = 0.5\rho v_b^2 \tag{4.10}$$

Here ρ is the air density, recommended value 1.25 kg/m^3, and v_b is the basic wind velocity, equal to 24 m/s (previously calculated). So,

$$q_b = 0.5 \times 1.25 \times v_b^2 = 0.5 \times 1.25 \times 24^2 = 360 \text{ N/m}^2$$

and

$$c_e(z) = q_p(z)/q_b$$

Therefore

$$q_p(z) = c_e(z)q_b$$

Referring to Fig. 2.1, for a structure of height 47 m from ground level, and terrain category II,

$$c_e(z) = 3.5$$

Therefore

$$\begin{aligned} q_p(z) &= \text{peak velocity pressure at 47 m height from ground level} \\ &= 3.5q_b = 3.5 \times 360 = 1260 \text{ N/m}^2 = 1.27 \text{ kN/m}^2 \end{aligned}$$

Wind pressures on surfaces based on wind tunnel experiments

The early wind tunnel experiments by Stanton (1908) on a model building provided pressure coefficients for the wind pressure distribution on vertical walls and roof slopes placed in the wind direction. The following external wind pressure coefficients c_{pe} for buildings with or without a roof slope were obtained:

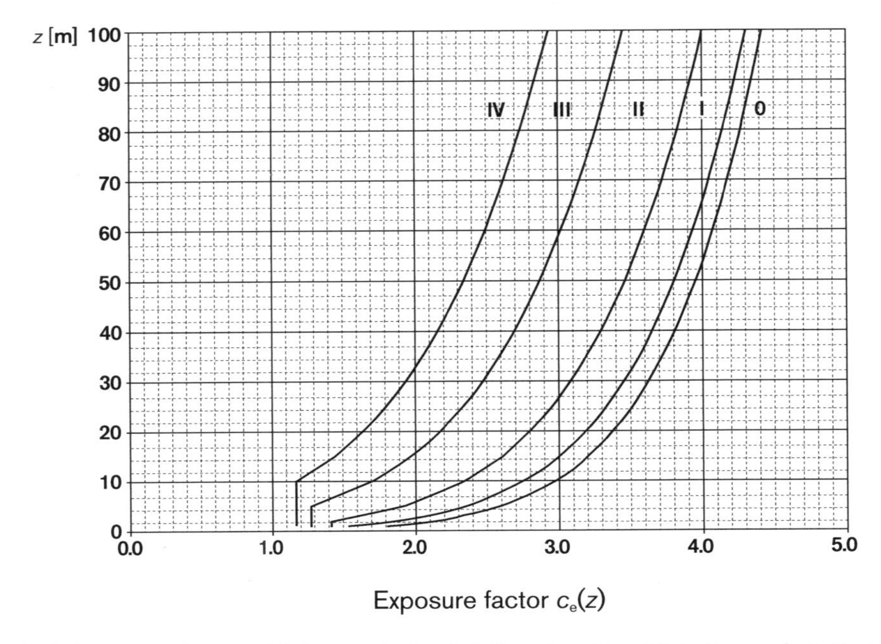

Fig. 2.1. Exposure factor $c_e(z)$ for $c_o = 0$, $k_r = 1.0$ (based on Fig. 4.2 in Eurocode 1, Part 1-4)

- *On windward vertical wall:* external pressure coefficient = c_{pe} = +0.5 (positive) (directed towards the surface).
- *On leeward vertical wall:* external suction coefficient = c_{pe} = −0.5 (negative) (directed away from the surface). In Holland, the values of the external pressure and suction coefficients were taken equal to +0.9 and −0.4, respectively.
- *On windward roof slope:* the results of the tunnel experiment also showed the following points. When the roof slope is 70° or more from the horizontal, the roof surface may be treated as equivalent to a vertical surface, and the external pressure coefficient c_{pe} is equal to +0.5 (positive). As the roof slope decreases, the positive normal wind pressure decreases. When the roof slope reaches 30°, the pressure reduces to zero. When the roof slope decreases below 30°, a negative normal pressure (suction) acts upwards normal to the slope. This suction pressure increases as the slope decreases and finally attains its full value when the slope reduces to zero (i.e. a flat roof). Thus expressions were found for the external pressure coefficients for roofs, as listed in Table 2.7. For example, if the roof slope is 45°, c_{pe} = (45/100 − 0.2) = +0.25. If the roof slope is 30°, c_{pe} = (30/60 − 0.5) = 0.0. If the roof slope is 10°, c_{pe} = (10/30 − 1.0) = −0.67 (upwards suction). If the roof slope is 0°, c_{pe} = (0/30 − 1) = −1.0 (upwards suction).

Table 2.7. Expressions for external pressure coefficients for roofs

θ (roof slope) (°)	Pressure coefficient normal to roof
45–70	θ/100–0.2
30–45	θ/60–0.5
0–30	θ/30–1.0

- *On leeward roof slope*, based on the same experiment results: for all roof slopes, $c_{pe} = -1.0$ (upwards suction).

Thus, for a flat roof, for the windward half, $c_{pe} = -1.0$ (upwards suction). For the leeward half, $c_{pe} = -0.5$ (upwards suction).

In addition to the external wind pressures on a building subjected to wind, a building is also subjected to internal pressures due to openings in the walls. Therefore, we have to also consider the internal pressure coefficients c_{pi}. When the wind blows into a building through an opening facing in the direction opposite to the wind blowing onto the building, the resultant effect is the development of internal pressure within the building.

Positive internal pressure: if wind blows into an open-sided building or through a large open door into a workshop, the internal pressure tries to force the roof and side coverings outwards and will cause a positive internal pressure.

Negative internal pressure: if wind blows in the opposite direction, tending to pull the roof and side coverings inwards, a negative internal pressure (suction) is created within the building.

In shops of normal permeability (covered with corrugated sheets), the coefficient of internal suction $c_{pi} = \pm 0.2$. In buildings with large openings (in the case of industrial buildings), the coefficient of internal suction $c_{pi} = \pm 0.5$.

A negative value implies internal suction, i.e. the inside pressure is away from the inner surfaces, and a positive value implies internal pressure, i.e. the inside pressure is towards the inner surfaces.

Wind pressures on surfaces, based on Eurocode 1, Part 1-4

First, we consider the external wind pressure coefficients for buildings with or without a roof. The external pressure acting on the external surfaces is given by the following equation:

$$w_e = q_p(z_e)c_{pe}$$

where w_e is the external pressure, $q_p(z_e)$ is the peak velocity pressure, z_e is the reference height for the external pressure and c_{pe} is the pressure coefficient for the external pressure. The value of c_{pe} depends on the ratio h/d for the building, where h is the total height of the building up to the apex in the case of a pitched roof, and d is the depth of the building.

For the external pressure coefficients on vertical faces, we refer to Table 2.8 and Fig. 2.2. Table 2.8 shows the recommended values of the external pressure coefficient for

Table 2.8. Recommended values of external pressure coefficient[a] c_{pe} for the vertical walls of a rectangular-plan building (based on Table 7.1 of Eurocode 1, Part 1-4)

	Zone				
h/d	A	B	C	D	E
5	−1.2	−0.8	−0.5	+0.8	−0.7
1	−1.2	−0.8	−0.5	+0.8	−0.5
≤0.25	−1.2	−0.8	−0.5	+0.7	−0.3

[a] The symbol used in the table in the Eurocode is $c_{pe,10}$, which appears to refer to wind tunnel measurements at a height of 10 m above ground level.

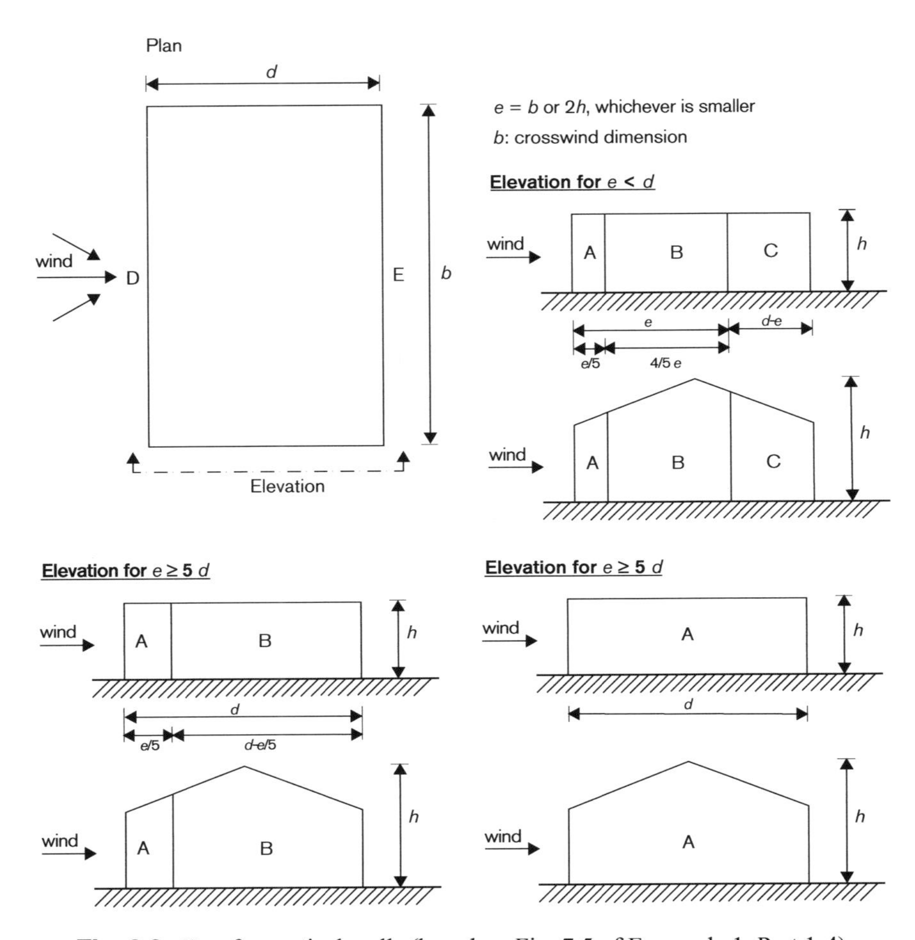

Fig. 2.2. Key for vertical walls (based on Fig. 7.5 of Eurocode 1, Part 1-4)

the vertical walls of a rectangular-plan building as shown in Fig. 2.2. For buildings with $h/d > 5$, the resulting force is multiplied by 1.0, and for $h/d < 1$, the resulting force is multiplied by 0.85. In Table 2.8, the windward face is denoted by zone D, the leeward face by zone E and the side faces by A, B and C. The values of $c_{pe,\,10}$ should be considered in design applications. For example, referring to Table 2.8, for a building with $h/d = 1$, the external pressure coefficient c_{pe} on the windward face D is +0.8, and the external pressure coefficient on the leeward face E is −0.5.

For the external pressure coefficients for duopitch roofs, on the windward and leeward slopes for roofs of various pitch angles, we refer to Table 7.4a of Eurocode 1, part 1-4 and Fig. 7.8 part 1-4, where the windward faces are denoted by F, G and H. The leeward faces are denoted by I and J. Various values of pressure coefficients are given for various pitch angles. For example, referring to Table 7.4a of Eurocode 1, Part 1-4, for a duopitch roof with pitch angle $\alpha = 15°$, the external pressure coefficient c_{pe} for the windward face H is −0.9, and the external pressure coefficient for the leeward face I is −0.5.

Now, we consider the internal pressure coefficients for buildings. Referring to Eurocode 1, Part 1-4, the wind pressure acting on the *internal* surfaces of a structure is expressed in the following form:

$$w_i = q_p(z_i)c_{pi} \tag{5.2}$$

where c_{pi} is the internal pressure coefficient. The Eurocode stipulates the following values of internal pressure coefficients:

- For a building where the area of the openings in the dominant face is twice the area of the openings in the remaining faces, the value of the coefficient is given by the equation

$$c_{pi} = 0.75c_{pe} \tag{7.1}$$

- For a building where the area of the openings in the dominant face is at least three times the area of the openings in the remaining faces, the value of the coefficient is given by the equation

$$c_{pi} = 0.9c_{pe} \tag{7.2}$$

- For buildings without a dominant face, the internal pressure coefficient should be determined from Fig. 7.13 of Eurocode 1, Part 1-4, and is a function of the ratio

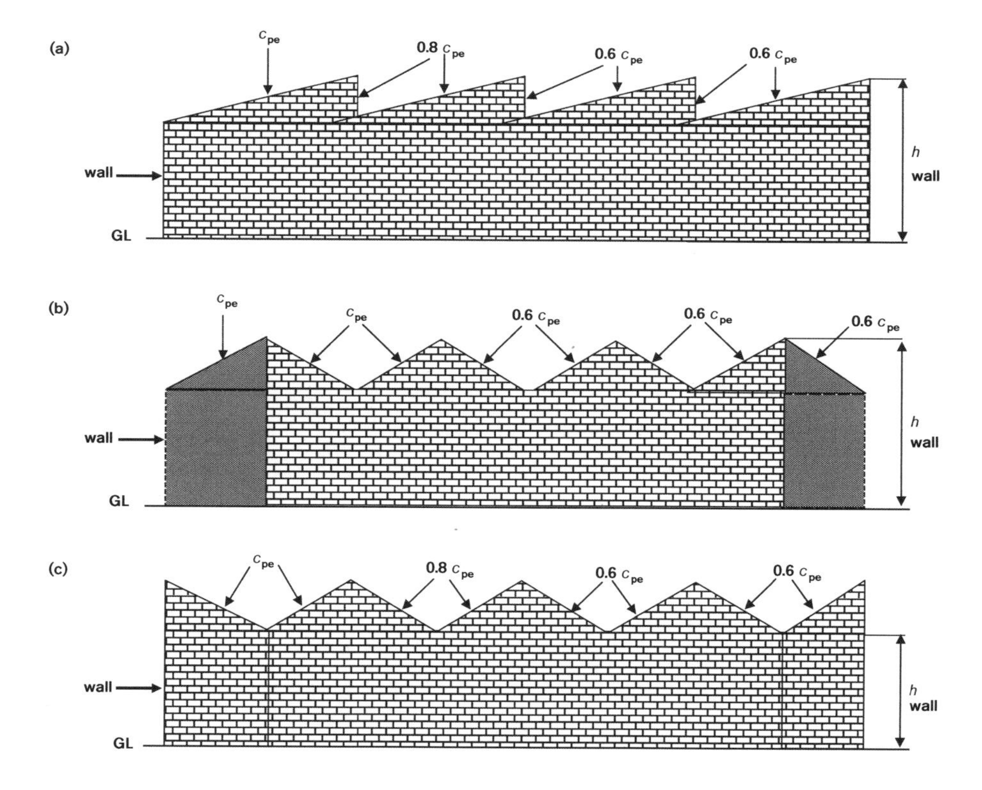

Fig. 2.3. Key for multispan roof

h/d and the opening ratio μ. Where it is not possible to calculate the value of μ, the following values of the internal pressure coefficient are justified: $c_{pi} = +0.2$ and $c_{pi} = -0.3$

For the external pressure coefficients for multispan duopitch roofs, we refer to Fig. 2.3(c) (based on Fig. 7.10(c) of Eurocode 1, Part 1-4). The first c_{pe} for the first roof slope is calculated as for a monopitch roof. The second and all following c_{pe}'s are calculated as for duopitch roofs.

2.3.1.5 Example 1

Calculate the pressure coefficients for a single-bay building with the following data:

- height of building = $h = 35.5$ m;
- depth of building = $d = 27.0$ m;
- roof slope = $\alpha = 6.28°$;
- wind blowing from right.

Based on Eurocode 1, Part 1-4, we calculate the coefficients as follows.

For the external wind pressure coefficients c_{pe}, we refer to Table 2.8, with $h/d = 1.37$. Therefore, on the windward vertical wall, $c_{pe} = +0.8$ (acting towards the wall, normal to the wall). On the leeward vertical wall, $c_{pe} = -0.5$ (suction, acting outwards normal to the wall). Referring to Table 7.4(a) of Eurocode 1, Part 1-4, for a roof pitch angle $\alpha = 6.28°$, on the windward roof slope, $c_{pe} = -0.6$ (acting outwards normal to the roof). On the leeward roof slope, $c_{pe} = -0.6$ (acting outwards normal to the roof) (see Fig. 2.4(a)).

For the internal pressure coefficients (suction) c_{pi} (the area of the openings in the dominant face is twice the area of the openings in the remaining faces, assumed to be of high permeability),

$$c_{pi} = 0.75c_{pe} \tag{7.1}$$

Therefore, on the windward vertical wall, $c_{pi} = -0.75c_{pe} = -0.75 \times 0.8 = -0.6$ (acting inwards, i.e. suction). On the leeward vertical wall, $c_{pi} = 0.75c_{pe} = -0.75 \times 0.5 = -0.4$ (acting inwards, i.e. suction). On the windward roof slope, $c_{pi} = 0.75c_{pe} = -0.75 \times 0.6 = -0.45$ (acting inwards, i.e. suction). On the leeward roof slope, $c_{pi} = -0.75c_{pe} = -0.75 \times 0.6 = -0.45$ (acting inwards, i.e. suction). (See Fig. 2.4(b).)

For the internal positive pressure coefficients c_{pi} (acting outwards from within the building), we have the following. On the windward vertical wall, $c_{pi} = +0.758 \times 0.8 = +0.6$. On the leeward vertical wall, $c_{pi} = +0.75 \times 0.5 = 0.4$. On the windward roof slope, $c_{pi} = +0.75 \times 0.6 = +0.45$. On the leeward roof slope, $c_{pi} = +0.75 \times 0.6 = +0.45$. (See Fig. 2.4(d).)

The resultant pressure coefficients ($c_{pe} - c_{pi}$) with an internal suction are:

- On windward vertical wall, $+0.8 - (-0.6) = +1.4 \rightarrow$
- On leeward vertical wall, $-0.5 - (-0.4) = -0.10 \rightarrow$
- On windward roof slope, $-0.6 - (-0.45) = -0.15 \uparrow$
- On leeward roof slope, $-0.6 - (-0.45) = -0.15 \uparrow$

(See Fig. 2.4(c).)

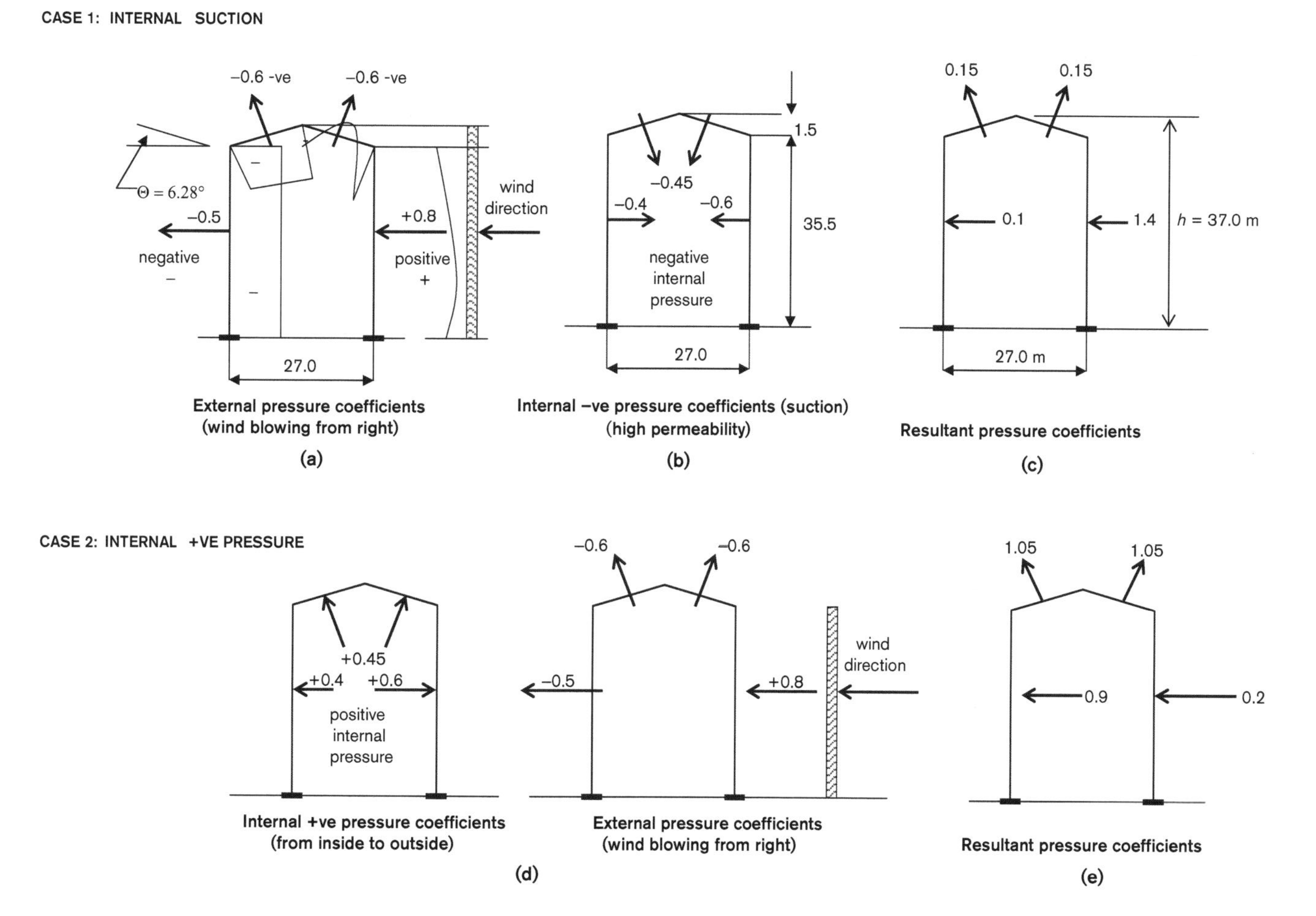

Fig. 2.4. Wind pressure coefficients for single-bay building

The resultant pressure coefficients ($c_{pe} - c_{pi}$) with an internal positive pressure are:

- On windward vertical wall, $+0.8 - (+0.6) = +0.20 \rightarrow$
- On leeward vertical wall, $-0.5 - (+0.4) = -0.90 \leftarrow$
- On windward roof slope, $-0.6 - (+0.45) = -1.05 \uparrow$
- On leeward roof slope, $-0.6 - (+0.45) = -1.05 \uparrow$

The values of the pressure coefficients are shown in Fig. 2.4(e).

2.3.1.6 Example 2

Consider our case, which is a multiple-bay building as shown in Figs 1.3 and 2.2:

- height of building in first, third and fourth bays from ground level = $h = 35.5$ m;
- height of building in second bay from ground level = $h = 45.5$ m;
- roof slope $\alpha = 6.28°$;
- depth of building $d = 96$ m.

From the above data, we arrive the wind pressure coefficients in accordance with Eurocode 1, Part 1-4, when the wind is blowing from right to left. Modifying factors for the external pressure coefficients on each span are derived from Clause 7.2.7 of the Eurocode for duopitch multispan roofs according to Fig. 2.3 (based on Fig. 7.10(c) of Eurocode 1, Part 1-4). Therefore, referring to equation (7.1) of the Eurocode with $h/d = 35.50/96 = 0.37 < 1$, we obtain the following results.

External pressure coefficients

On the vertical walls:

external pressure coefficient on windward wall $c_{pe} = +0.8$.
external pressure coefficient on leeward wall $c_{pe} = -0.5$.

On the duopitch roofs, referring to Table 7.4(a) of Eurocode 1, Part 1-4 (external pressure coefficients for duopitch roofs), with a roof pitch angle $\alpha = 6.28°$, we have the following:

- On the roof of the first bay,
  external pressure coefficient on windward slope $c_{pe} = -0.6$,
  external pressure coefficient on leeward slope $c_{pe} = -0.6$.

- On the roof of the second bay,
  external pressure coefficient on windward slope $c_{pe} = -0.6$,
  external pressure coefficient on leeward slope $c_{pe} = -0.6$.

As the second roof is higher than the first roof, it is not sheltered. So no reduction factor should be considered for it.

- On the roofs of the third and fourth bays,
  external pressure coefficient on windward slope = $0.6c_{pe} = 0.6 \times (-0.6) = -0.4$,
  external pressure coefficient on leeward slope = $0.6c_{pe} = 0.6 \times (-0.6) = -0.4$

(where 0.6 is the reduction factor; see Fig. 7.10(c) of Eurocode 1, Part 1-4).

Internal pressure coefficients c_{pi} (suction)

Referring to Clause 7.2.9, Note 2 (internal pressure), where is not possible to calculate exactly the areas of the openings in the surfaces, then the values of c_{pi} should be taken equal to +0.2 and −0.3 (suction). Therefore

internal pressure coefficients on all spans = −0.3.

Resultant pressure coefficients (c_{pe} − c_{pi})

On windward vertical wall, $c_{pe} - c_{pi} = 0.8 - (-0.3) = +1.1$.
On leeward vertical wall = −0.5 − (−0.3) = −0.20.
On first windward roof slope = −0.6 − (−0.3) = −0.30.
On first leeward roof slope = −0.6 − (−0.3) = −0.30.
On second windward roof slope = −0.6 − (0.3) = −0.30.
On second leeward roof slope = −0.6 − (−0.3) = −0.30.
On third and fourth windward roof slopes = −0.4 − (−0.3) = −0.10.
On third and fourth leeward roof slopes = −0.4 − (−0.3) = −0.10.

The values of the pressure coefficients are shown in Fig. 2.5.

2.4 Properties and specification of materials

This is done based on Eurocode 3, Part 1-1, BS EN 1993-1-1: 2005 (Eurocode, 2005a). The steel structure is designed using the ULS method. So, the structural strength is based on the yield strength of structural steel f_y. We describe first the properties and strength of structural steel and fasteners (bolts and welds).

2.4.1 Properties and strength of structural steel and fasteners

2.4.1.1 Type of construction material

We refer to Table 3.1 of Annex A of Eurocode 3, Part 1-1, reproduced in Appendix B of this book. The structural steel is of grade S 275 (EN 10025-2): $f_y = 275$ N/mm^2 for $t \leq 40$ mm, and $f_y = 255$ N/mm^2 for $t \leq 80$ mm, where t is the nominal thickness of the element.

2.4.1.2 Design strength (f_y)

The design strength of weldable structural steel should conform to the grades and product standards specified in BS EN 100251: 2004 (British Standards Institution, 2004) (see BS EN 1993-1-8: 2005(E) (Eurocode, 2005c)). The design strength f_y should be taken equal to the minimum yield strength, and the ultimate tensile strength f_u should be taken as 430 N/mm^2.

The above code stipulates values of the design strength of different grades of steel for various thicknesses of the product. For example, for steel grade S 275 and thickness 40 mm or less, $f_y = 275$ N/mm^2. With an increase in thickness, the value of f_y decreases. For steel grade S 355 and thickness 40 mm or less, $f_y = 355$ N/mm^2, and f_y decreases as the thickness increases. For steel grade S 450 and thickness 40 mm or less, $f_y = 440$ N/mm^2, and f_y again decreases as the thickness increases.

For details, refer to Table 3.1 of Annex A of Eurocode 3, Part 1-1.

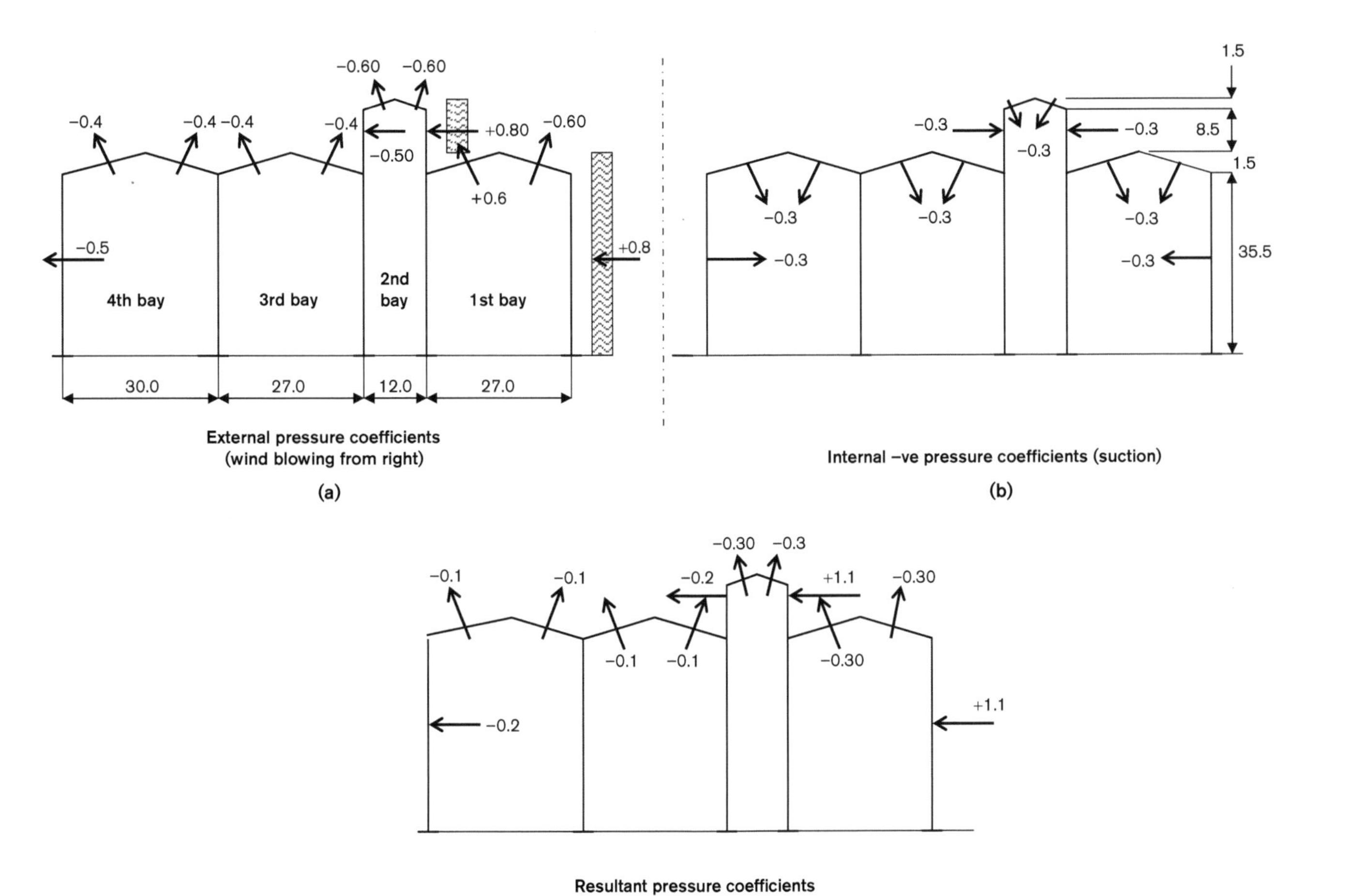

Fig. 2.5. Wind pressure coefficients on multispan building (wind blowing from right)

2.4.1.3 Elastic properties of steel

We refer to Clause 3.2.6, "Design values of material coefficients", of Eurocode 3, Part 1-1. The values of the elastic properties of steel are given there as follows:

Modulus of elasticity $E = 210\,000$ N/mm^2.

Shear modulus $G = E/[2(1 + \upsilon)] = 810\,000$ N/mm^2.

Poisson's ratio $\nu = 0.30$.

Coefficient of linear thermal expansion $\alpha = 12 \times 10^{-6}$ per °C

(in the ambient temperature range).

2.4.1.4. Bolts and welded connections, based on Eurocode 3, Part 1-8 (Eurocode, 2005c)

The specifications for the strengths and properties of fasteners (bolts, nuts and washers, including friction grip bolts) are given in Eurocode 3, Part 1-8, BS EN 1993-1-8: 2005.

2.4.1.5 Welding consumables

All welding consumables, including covered electrodes, wires, filler rods, fluxes and shielding gases, should conform to BS EN 1993-1-8: 2005.

2.4.2 Partial factors γ_M of resistance in the ultimate-limit-state concept

In the ULS concept, the structure should be designed to a limiting stage beyond which the structure becomes unfit for its intended use. Referring to Clause 6.1 of Eurocode 3, Part 1-1, the following recommended values of partial factors γ_M should be applied to the various characteristic values of resistance:

- resistance of cross-sections, whatever the class is: $\gamma_{M0} = 1.00$;
- resistance of members to instability, assessed by member checks: $\gamma_{M1} = 1.00$;
- resistance of cross-sections in tension to fracture: $\gamma_{M2} = 1.25$;
- resistance of joints: see BS EN 1993-1-8: 2005.

2.4.3 Ultimate limit state

Ultimate limit states relate to the safety of a structure as a whole or of part of it. In checking the strength of a structure or of any part of it, the specified loads should generally be multiplied by the relevant partial factors. Thus, γ_G is the partial factor for permanent loads, and γ_Q is the partial factor for variable loads. The resulting factored loads should be applied in the most unfavourable combination so that the load-carrying capacity of the members sustains adequate strength without allowing any collapse.

The method of design based on the above ultimate limit states is known as the *ultimate-limit-state (ULS) method.*

2.4.4 Serviceability limit state

Serviceability limit states define the limit beyond which the specified service conditions are no longer fulfilled. In serviceability limit states, the specified loads are generally unfactored except in the case of combinations of imposed loads and wind loads, in which case 80% of the full specified load is taken into account.

The method of design based on the above serviceability limit states is called the *serviceability limit state (SLS) method.* This method is applied in checking deflection, vibration etc.

2.4.5 Load combinations

In structural design for load combinations using the ULS and SLS methods, the partial factors γ_G for permanent actions and partial factors for variable actions should be taken from Table A1.2(B) of BS EN 1990: 2002(E) (see Appendix B).

2.5 Specifications for selecting the structural components

2.5.1 Length of span

In the design, the effective spans of beams, girders and roof trusses should be taken as equal to the centre-to-centre distance between the end bearings, or the distance from end to end when the beam, girder or roof truss is connected between stanchions or girders by cleats.

2.5.2 Roof trusses

The following specifications should apply in selecting the spacing, pitch and form of roof trusses.

2.5.2.1 Spacing

In general, the spacing of frames is constant in normal buildings. But in industrial buildings, the spacing of frames may vary to accommodate machines and equipment.

The spacing of trusses should be limited so that an economical section can be designed and used for the purlins. As a guide, the following spacings should be applied for various spans:

- For spans up to 18 m, the spacing should preferably be 4 m but should not exceed 5 m.
- For spans from 18 to 25 m, the spacing should not exceed 6 m.

In our case, the spacing of trusses is limited to 6.0 m.

2.5.2.2 Pitch

Pitch of roof = (height of apex from centre of span)/span.
Normally, the pitch should preferably be not less than 1/4 (a slope of 1 in 2), and in no case less than 1/5.

However, in industrial structures, the pitch may be lower than the above ratios, say 1/8 or even less. Such a low-pitch roof allows one to develop a rigid connection between the roof trusses and stanchions to resist moments at the connections due to high crane surges and wind forces. In the case of wind and snow loadings, this low pitch may increase the stresses in the members.

In our case, a pitch of 1/8 is adopted.

2.5.2.3 Form

In general, rafters are designed to resist direct stress. So the rafters should preferably be equally subdivided so that the purlins are located at panel (node) points to avoid the development of any additional bending stress. When it is not practically possible to locate all of the purlins at node points, the rafters should be designed as a beam subjected to combined direct and bending stresses.

In our case, all the purlins are located at node points of the rafters, with equal spacing.

2.6 Conventions for member axes

We refer to Clause 1.7, "Conventions for member axes", and Fig. 1.1 of Eurocode 3, Part 1-1.

The conventions for steel members of I section are:

- *x–x* axis along the member;
- *y–y* cross-section axis parallel to the flanges;
- *z–z* cross-section axis perpendicular to the flanges.

The conventions for steel members of angle section are:

- *y–y* axis parallel to the smaller leg;
- *z–z* axis perpendicular to the smaller leg.

2.7 Model design of beam and column using Eurocode 3 and BS 5950, and comparison of the results

2.7.1 Model design of a beam

Example 1. To design a beam for use in a multibay structural-steel-framed building subjected to loadings (dead, imposed, wind and crane). The beam supports a reinforced concrete slab 200 mm thick. The ends of the beam are connected to columns, and the end conditions are considered continuous (fixed) (see Fig. 2.6).

Effective span of beam $L = 11.75$ m

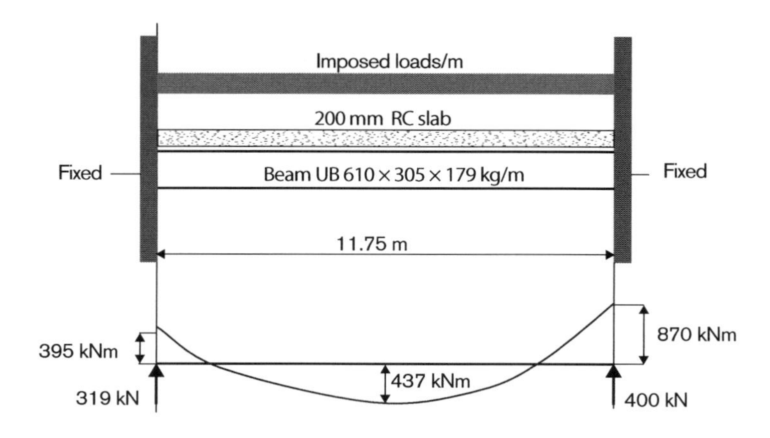

Fig. 2.6. Beam with loads and bending-moment diagram

Analyses were carried out with various types of loadings, and the results of the analyses gave the following values:

Ultimate design moment at left support, a, $M_{Ed,a} = 395$ kN m
Ultimate design moment at right support, c, $M_{Ed,c} = 870$ kN m $= M_u$
Ultimate design span moment at b, $M_{Ed,b} = 437$ kN m
Ultimate shear at right support, c, $V_{Ed,c} = 400$ kN $= F_v$

2.7.1.1 To design the section of the beam in combined bending and shear, based on Eurocode 3

Initial sizing of section

Try a section UB610 × 305 × 179 kg/m; grade of steel S 275; $f_y = 275$ N/mm^2. The properties of the section are as follows.

Depth of section $h = 620.2$ mm.
Depth between fillets $h_w = 540$ mm.
Width of flange $b = 307.1$ mm.
Thickness of web $t_w = 14.1$ mm.
Thickness of flange $t_f = 23.6$ mm.
Root radius $r = 16.1$ mm.
Radius of gyration, y–y axis, $i_y = 25.9$ cm.
Radius of gyration, z–z axis, $i_z = 7.07$ cm.
Elastic modulus, y–y axis, $W_y = 4940$ cm^3.
Elastic modulus, z–z axis, $W_z = 743$ cm^3.
Plastic modulus, y–y axis, $W_{pl,y} = 5550$ cm^3.
Plastic modulus, z–z axis, $W_{pl,z} = 1140$ cm^3.
Area of section $A = 228$ cm^2.

Section classification (see Clause 5.5 and Annex A)

Flange

Stress factor $\varepsilon = (235/f_y)^{0.5} = (235/275)^{0.5} = 0.92$
Outstand of flange $c = (b - t_w - 2r)/2 = (307.1 - 14.1 - 2 \times 16.1)/2 = 130.4$
Ratio $c/t_f = 130.4/23.6 = 5.52$
$9\varepsilon = 9 \times 0.92 = 8.28$

For class 1 section classification, the limiting value of $c/t_f \leq 9\varepsilon$. In the present case we have

$5.52 \leq 8.28$.

So, the flange satisfies the conditions for class 1 section classification.

Web

Ratio $h_w/t_w = 540/14.1 = 38.29$; $72\varepsilon = 72 \times 0.92 = 66.24$

referring to Table 5.2 (sheet 1 of 3) of Eurocode 3, Part 1-1 (see Appendix B), as the web is subjected to bending only. Therefore

$h/t_w \leq 72\varepsilon$

For class 1 section classification, the limiting value of h_w/t_w (38.9) < 72ε (66).

So, the web satisfies the conditions for class 1 section classification.

Moment capacity

In accordance with Clause 6.2.8 of Eurocode 3, where bending and shear act simultaneously on a structural member, the moment capacity should be calculated in the following way:

- where a shear force is present, allowance should be made for its effect on the moment resistance;
- where the shear force is less than half the plastic shear resistance, its effect on the moment resistance may be neglected;
- otherwise, the reduced moment resistance should be taken as the design resistance of the section, calculated using a reduced yield strength $(1 - \rho)f_y$, where

$\rho = ((2V_{Ed}/V_{pl,Rd}) - 1)^2$ and $V_{pl,Rd} = A_v(f_y/\sqrt{3})/\gamma_{M0}$

where A_v is the shear area, equal to $A - 2bt_f + 2(t_w + 2r)t_f = 10\,528$ mm^2, but which should not be less than $h_w t_w = 7614$ mm^2; this condition is satisfied. γ_{M0} is the partial safety factor for resistance of the cross-section, equal to 1.0 (see Clause 6.1 of Eurocode 3).

Therefore, in our case,

$V_{pl,Rd} = (10528 \times (275/\sqrt{3})/1.0)/10^3 = 1672$ kN

and

$V_{pl,Rd}/2 = 836$ kN

So,

V_{Ed} (400 kN) < $V_{pl,Rd}/2$ (836 kN)

So, the effect of shear force on the reduction of plastic resistance moment need not be considered. Therefore,

plastic moment capacity $M_{pl,Rd} = 275 \times 5550 \times 10^3/10^6$
$= 1526$ kN m > 870 kN m   OK

Shear buckling resistance

Shear buckling resistance need not be checked if the ratio $h_w/t_w < 72\varepsilon$. In our case,

$h_w/t_w = 38.9 < 72\varepsilon$ (66.24)

Therefore, shear buckling resistance need not be checked.

Buckling resistance moment

Referring to Fig. 2.6, the beam is fixed at both ends. The top flange is restrained by a reinforced concrete slab (because sufficient friction is developed between the steel and concrete surfaces), but the bottom compression flange near the supports is unrestrained against lateral buckling. The unsupported unrestrained length of the compression zone of the bottom flange near the right support is 2.75 m from the point of the end restraint to the point of contraflexure (obtained from the combined moment diagram). The effective unsupported length of the compression flange against buckling, L_c, is $0.85 \times 2.75 = 2.3$ m. The buckling resistance moment along the major axis of the member is calculated in the following way, referring to Clause 6.3.2.4 of Eurocode 3, "Simplified assessment methods for beams with restraints in buckling".

Members with lateral resistance to the compression flange are not susceptible to lateral–torsional buckling if the length L_c between restraints or the resulting slenderness $\bar{\lambda}_f$ of the equivalent compression flange satisfies

$$\bar{\lambda}_f = (k_c L_c)/(i_{f,z}\lambda_1) \leq \bar{\lambda}_{c0} M_{c,Rd}/My_{,Ed}$$

where k_c is the slenderness correction factor for the moment distribution with one-end restraint conditions.

In our case, one end is fixed and the other end is at the point of contraflexure, and so $k_c = 0.82$ (see Table 6.6 of Eurocode 3, Part 1-1).

$My_{,Ed} = 870$ kN m (ultimate design moment)
$M_{c,Rd} = W_y f_y/\gamma_{M1} = 5550 \times 10^3 \times 275/10^6 = 1526$ kN m

where $\gamma_{M1} = 1.0$
$i_{f,z}$ = radius of gyration of equivalent compression flange about minor axis = 8.11 cm
$\lambda_1 = 93.9\varepsilon = 86.4$
$\bar{\lambda}_{c0}$ = slenderness limit of the equivalent compression flange = $\lambda_{LT,0} + 0.1 = 0.4 + 0.1 = 0.5$

(see Clause 6.3.2.3 of Eurocode 3 for the value of $\lambda_{LT,0}$).

L_c = effective length between restraints = 2.3 m (already calculated)
$\bar{\lambda}_f = (k_c L_c)/(i_{f,z}\lambda_1) = (0.82 \times 230)/(8.11 \times 86.4) = 0.27$

and

$$\bar{\lambda}_{c0} M_{c,Rd}/My_{,Ed} = 0.5 \times 1526/870 = 0.87$$

Therefore, there is no necessity for a reduction of the design buckling resistance moment, and $My_{,Ed}/M_{b,Rd} = 870/1526 = 0.57 < 1$ (see equation (6.54) of Eurocode 3).

Therefore we adopt UB610 × 305 × 179 kg/m for the beam.

2.7.1.2 To design the section of the beam in moment and shear (based on BS 5950: 2000-1) (British Standards Institution, 2000)

Ultimate design moment at left support, a, $M_{Ed,a} = 395$ kN m

Ultimate design moment at right support, c, $M_{Ed,c} = 870 \text{ kN m} = M_u$

Ultimate design span moment at b, $M_{\text{Ed,b}} = 437$ kN m

Ultimate shear at right support, c, $V_{\text{Ed,c}} = 400$ kN $= F_v$

Initial sizing of section

Try a section UB610 × 305 × 179 kg/m; grade of steel S 275; $p_y = 275$ N/mm^2. The properties of the section are as follows.

Depth of section $D = 620.2$ mm

Depth between fillets $d = 540$ mm

Width of flange $B = 307.1$ mm

Thickness of web $t = 14.1$ mm

Thickness of flange $T = 23.6$ mm

Root radius $r = 16.5$ mm

Radius of gyration, x–x axis, $r_x = 25.9$ cm

Radius of gyration, y–y axis, $r_y = 7.07$ cm

Elastic modulus, x–x axis, $Z_x = 4940$ cm^3

Elastic modulus, y–y axis, $Z_y = 743$ cm^3

Plastic modulus, x–x axis, $S_x = 5550$ cm^3

Plastic modulus, y–y axis, $S_y = 1140$ cm^3

Moment of inertia, x–x axis, $I_x = 153\,000$ cm^4

Area of section $A = 228$ cm^2

$b = B/2 = 307.1/2 = 153.6$

$b/T = 153.6/23.6 = 6.51$, $d/t = 38.3$ and $D/T = 620.2/23.6 = 26.3$

Section classification

Firstly, before designing the section, we have to classify the section into one of the following classes:

- *Class 1, plastic:* cross-section with hinge rotation capacity.
- *Class 2, compact:* cross-section with plastic moment capacity.
- *Class 3, semi-compact:* cross-section in which the stress in the extreme compression fibre can reach the design strength but a plastic moment capacity cannot be developed.
- *Class 4, slender:* cross-section for which we have to have a special allowance owing to the effects of local bending.

Let us consider our assumed section UB610 × 305 × 179 kg/m; grade of steel S 275. Referring to Table 9 of BS 5950: 2000-1 ("Design strength p_y"), assuming steel grade S 275,

thickness of flange $T = 23.6$ mm

Since the thickness of the flange is greater than 16 mm and less than 40 mm, the design strength $p_y = 265$ N/mm^2. Referring to Table 11 of BS 5950: 2000-1 ("Limiting width-to-thickness ratio for section other than CHS and RHS"),

$\varepsilon = (275/p_y)^{0.5} = (275/265)^{0.5} = 1.02$

For class 1, plastic, the limiting value of $9\varepsilon \geq b/T$ (6.51). In our case,

$9 \times 1.02 = 9.2 > 6.51$

Also, for a web with its central axis at mid depth, for a class 1 plastic section the limiting value of $80\varepsilon \geq d/t$ (38.3). In our case,

$80 \times 1.02 > 38.3$

Therefore the section chosen is a class 1 plastic section.

To check shear capacity of section

Referring to Clause 4.2.3 ("Shear capacity"), the shear force F_v should not be greater than the shear capacity P_v of the section. In our case,

$P_v = 0.6p_yA_v$

where A_v = shear area = $tD = 14.1 \times 620.2 = 8745\ \text{mm}^2$, and so

$P_v = 0.6 \times 265 \times 8745/10^3 = 1390\ \text{kN} > F_v\ (400\ \text{kN})$

Since d/t (38.3) < 70ε (71.4), the section need not be checked for shear buckling.

To check moment capacity of section

Referring to Clause 4.2.5.2 ("Moment capacity with low shear"), if the shear force F_v does not exceed 60% of the shear capacity P_v, then, for class 1 and class 2 sections, the moment capacity $M_c = p_yS_x$. In our case, $F_v = 400$ kN, and 60% of $P_v = 0.6 \times 1390 = 834$ kN. So, $F_v < 0.6P_v$. Therefore

$$M_c = p_yS_x < 1.2p_yZ_x = 265 \times 5550/10^3 < 1.2 \times 265 \times 4940/10^3$$
$$= 1471 < 1571$$

We adopt the lower value.

Therefore the ultimate design moment M_u (870 kN m) < M_c (1471). Satisfactory

To check for lateral–torsional buckling

Referring to Clause 4.3 ("Lateral–torsional buckling"), as already discussed, the effective length of the unstrained bottom compression flange L_y is 2.3 m. Referring to Clause 4.3.6.7 ("Equivalent slenderness λ_{LT}"),

effective slenderness $\lambda_{LT} = \lambda_y uv\beta_w{}^{0.5}$

where λy = y–y axis slenderness = $L_y/r_y = 230/7.07 = 33$, and u is the buckling parameter. Referring to Clause 4.3.6.8 ("Buckling parameter u and torsional index x"), for rolled I and H sections or channels with equal flanges, and with $x = D/T = 20.4$, $u = 0.9$. v is the slenderness factor for the beam. With $D/T = 20.4$ and $\lambda_y/x = 33/20.4 = 1.6$, referring to Table 19 of BS 5950: 2000-1, $v = 0.97$. β_w is a ratio; referring to Clause 4.3.6.9, for class 1 and class 2 sections, $\beta_w = 1.0$. Therefore

effective slenderness $\lambda_{LT} = \lambda_y uv\beta_w{}^{0.5} = 33 \times 0.9 \times 0.97 \times 1.0 = 28$

Referring to Table 16 ("Bending strength p_b for rolled section"), and with $\lambda_{LT} = 28$ and $p_y = 265$,

allowed bending stress $p_b = 265$ N/mm^2

Therefore

buckling moment of resistance $M_b = p_b S_x = 265 \times 5550/10^3 = 1471$ kN m $> M_u$ (870 kN m).

Ratio = 870/1471 = 0.59

Thus, when the results obtained by applying Eurocode 3 and BS 5950 are compared, Eurocode 3 gives a slightly higher value of the buckling moment of resistance.

2.7.2 Model design of a column

Example 2. To design a column of a structural-steel portal-framed building subjected to dead, imposed and wind loadings. The outside face of the column supports the side rails of the covering. The top end of the column is considered continuous (fixed) and the base is assumed hinged. The effective span of the column parallel to the major axis, L_y, is 11.8 m (see Fig. 2.7).

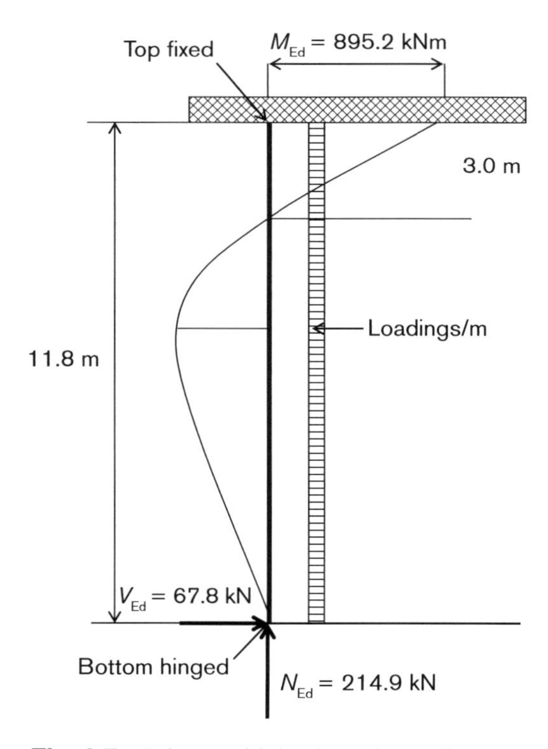

Fig. 2.7. Column with loads, and BM diagram

Analyses were carried out with various types of loadings, and the results gave the following values:

Ultimate design moment at top support $M_{Ed} = 895.2$ kN m.

Ultimate design shear $V_{Ed,c} = 67.8$ kN.

Ultimate design thrust $N_{Ed} = 214.9$ kN.

2.7.2.1 To design the section of a column in combined bending, shear and thrust, based on Eurocode 3

Initial sizing of section

Try a section UB610 × 305 × 179 kg/m; grade of steel S 275; $m_{fy} = 275$ N/mm^2. The properties of the section are as follows.

Depth of section $h = 620.2$ mm.

Depth between fillets $h_w = 540$ mm.

Width of flange $b = 307.1$ mm.

Thickness of web $t_w = 14.1$ mm.

Thickness of flange $t_f = 23.6$ mm.

Root radius $r = 16.1$ mm.

Radius of gyration, y–y axis, $i_y = 25.9$ cm.

Radius of gyration z–z axis, $i_z = 7.07$ cm.

Elastic modulus, y–y axis, $W_y = 4940$ cm^3.

Elastic modulus, z–z axis, $W_z = 743$ cm^3.

Plastic modulus, y–y axis, $W_{pl,y} = 5550$ cm^3.

Plastic modulus, z–z axis, $W_{pl,z} = 1140$ cm^3.

Area of section $A = 228$ cm^2.

Section classification

Flange

Stress factor $\varepsilon = (235/f_y)^{0.5} = 0.92$.

Outstand of flange $c = 130.4$ mm (previously calculated).

Ratio $c/t_f = 5.52$.

$9\varepsilon = 8.28$.

For class 1 section classification, the limiting value of $c/t_f \leq 9\varepsilon$. In the present case we have

$5.52 < 8.28$.

So, the flange satisfies the conditions for class 1 section classification.

Web

Ratio $h_w/t_w = 540/14.1 = 38.29$.

Referring to Table 5.2 (sheet 1) of Eurocode 3, Part 1-1, as the web is subjected to bending and compression, and assuming $\alpha \geq 0.5$, $c/t_w \leq 396\varepsilon/(13\alpha - 1) \leq 66$. For class 1 section classification, the limiting value of $h_w/t_w \leq 66$; in the present case,

$$38.29 < 66.$$

So, the web satisfies the conditions for class 1 section classification.

Moment capacity

In accordance with Clause 6.2.10 of Eurocode 3, where bending, shear and axial thrust act simultaneously on a structural member, the moment capacity is calculated in the following way:

- where shear and axial forces are present, allowance should be made for the effect of both the shear force and the axial force on the resistance moment.
- provided that the design value of the shear force V_{Ed} does not exceed 50% of the design plastic shear resistance $V_{pl,Rd}$, no reduction of the resistance defined for an axial force in Clause 6.2.9 is made, except where shear buckling reduces the section resistance.
- where V_{Ed} exceeds 50% of $V_{pl,Rd}$, the design resistance of the cross-section to the combination of a moment and an axial force should be calculated using the reduced yield strength $(1 - \rho)f_y$ of the shear area, where $\rho = (2V_{Ed}/V_{pl,Rd})^2$ and $V_{pl,Rd} = A_v f_y/(\sqrt{3}/\gamma_{M0})$.

When $h_w/t_w \leq 72\varepsilon$, as is required for a class 1 section classification, it should be assumed that the web is not susceptible to buckling, and the moment capacity should be calculated from the equation $My_{Rd} = f_y Wy$, provided the shear force $V_{Ed} < V_{pl,Rd}$. In our case, we have assumed a section in which $h_w/t_w \leq 72\varepsilon$. So, the web is not susceptible to buckling.

When the ultimate shear force $V_{Ed} \leq 0.5V_{pl,Rd}$, the ultimate shear force V_{Ed} is 67.8 kN m, and the design plastic shear capacity of the section is

$$V_{pl,Rd} = A_v(f_y/\sqrt{3})/\gamma_{M0},$$

where $A_v = 10\,528$ mm^2 and $\gamma_{M0} = 1.0$. So, in our case,

$$V_{pl,Rd} = (10\,528 \times (275/\sqrt{3})/1.0)/10^3 = 1672 \text{ kN m}$$

and $0.5V_{pl,Rd} = 836$ kN. So,

$$V_{Ed}\ (67.8) < 0.5V_{pl,Rd}.$$

Therefore the effect of shear force on the reduction of plastic resistance need not be considered.

When the ultimate axial force $N_{Ed} \leq 0.25N_{pl,Rd}$, the effect on the plastic resistance need not be considered. Here,

$$0.25N_{pl,Rd} = 0.25Af_y/\gamma_{M0} = 0.25 \times 228 \times 10^2 \times 275/10^3 = 1568 \text{ kN}.$$

In our case, N_{Ed} (214.9) < $0.25N_{pl,Rd}$ (1568 kN). So, the effect on the plastic resistance need not be considered. Therefore

$$\text{plastic moment capacity } M_{pl,Rd} = f_y W_{pl,y} = 275 \times 5550/10^3 = 1526 \text{ kN m} > 895.2 \text{ kN m}.$$

The result is satisfactory.

Shear buckling resistance

Shear buckling resistance need not be checked if the ratio $h_w/t_w \leq 396\varepsilon/(13\alpha - 1)$. In our case, $h_w/t_w = 38.29$ and $396\varepsilon/(13\alpha - 1) = 66$. So, $h_w/t_w \leq 396\varepsilon/(13\alpha - 1)$.

Therefore shear buckling resistance need not be checked.

Buckling resistance to compression

Ultimate design compression $N_{Ed} = 214.9$ kN.

Buckling resistance to compression $N_{b,Rd} = \chi A f_y/\gamma_{M1}$,

where

$$\chi = 1/[\Phi + (\Phi^2 - \bar{\lambda}^2)^{0.5}]$$
$$\Phi = 0.5[1 + \alpha(\bar{\lambda} - 0.2) + \bar{\lambda}^2]$$

$\bar{\lambda} = L_{cr}/(i\lambda_1)$,
and where L_{cry} = buckling resistance for major axis = $0.85 \times (15 - 3.2) = 10$ m (the factor of 0.85 is taken because the bottom is assumed hinged and the top is assumed fixed), $\lambda_{1y} = 93.9\varepsilon = 93.9 \times 0.92 = 86.4$ and $i_y = 25.9$ cm. Therefore

$$\bar{\lambda}_y = 1000/(25.9 \times 86.4) = 0.45$$

For the minor axis, $L_{crz} = 1.0 \times 200$ cm $= 200$ cm (as the column supports side rails at 200 cm spacing), $\lambda_1 = 86.4$ and $i_z = 7.07$ cm. Therefore

$$\bar{\lambda}_z = 200/(7.07 \times 86.4) = 0.32$$

and

$$h/b = 620/307.1 = 2 > 1.2$$

Referring to Table 6.2 of Eurocode 3, Part 1-1, and following the buckling curves "a" for buckling about the y–y axis and "b" for buckling about the z–z axis in Fig. 6.4 of that Eurocode,

reduction factor χ_y for major axis = 0.97

reduction factor χ_z for minor axis = 0.95

Therefore, the buckling resistance to compression is

$$N_{b,Rd} = \chi A f_y/\gamma_{M1} = 0.95 \times 228 \times 10^2 \times 275/10^3 = 5957 \text{ kN} > N_{Ed} \text{ (214.9 kN)}$$

Satisfactory

Buckling resistance moment

The column is assumed to be hinged at the base and fixed at the top. The outer flange, connected to the side rails at 2 m intervals, is restrained against buckling with an unrestrained buckling length of only 2.0 m, when subjected to the worst combinations of loadings (as previously shown). But the inner compression flange is not restrained near the top for a length of approximately 3 m at the point of contraflexure. So the unrestrained length is equal to 3.0 m from the top.

The buckling resistance moment can be calculated in the following simple way. Referring to Clause 6.3.2.4 of Eurocode 3 ("Simplified assessment methods for members with restraints in buckling"), members with lateral resistance to the compression flange are not susceptible to lateral–torsional buckling if the length L_c between restraints or the resulting slenderness $\bar{\lambda}_f$ of the equivalent compression flange satisfies

$$\bar{\lambda}_f = (k_c L_c)/(i_{f,z}\lambda_1) \leq \bar{\lambda}_{c0} M_{c,Rd}/M_{y,Ed}$$

where

k_c = slenderness correction factor for moment distribution between restraints = 0.82

(see Table 6.6 of Eurocode 3, Part 1-1);

$M_{y,Ed}$ = 895.2 kN m;

$M_{c,Rd} = W_y f_y/\gamma_{M1} = 5550 \times 10^3 \times 275/10^6 = 1526$ kN m, where $\gamma_{M1} = 1.0$;

$i_{f,z}$ = radius of gyration of equivalent compression flange about minor axis = 8.11 cm;

λ_1 = 86.4 (previously calculated);

$\bar{\lambda}_{c0}$ = slenderness limit of equivalent compression flange
$= \bar{\lambda}_{LT,0} + 0.1 = 0.4 + 0.1 = 0.5$

(see Clause 6.3.2.3 of Eurocode 3);

L_c = length between restraints = 300 cm;

$\bar{\lambda}_f = (k_c L_c)/(i_{f,z}\lambda_1) = (0.82 \times 300)/(8.11 \times 86.4) = 0.35$;

$\bar{\lambda}_{c0} M_{c,Rd}/M_{y,Ed} = 0.5 \times 1526/895.2 = 0.85 > 0.35$.

So, there is no necessity for a reduction of the design buckling resistance moment, and

$$M_{Ed}/M_{b,Rd} + N_{Ed}/N_{b,Rd} = 895.2/1526 + 214.9/5330 = 0.59 + 0.04 = 0.63 < 1.0.$$

Satisfactory

Therefore we adopt UB610 × 305 × 9 kg/m for the column.

2.7.2.2 To design the section in combined bending, shear and thrust (based on BS 5950: 2000)

Ultimate design moment at top support M_{Ed} = 895.2 kN m.

Ultimate design shear $V_{Ed,c}$ = 67.8 kN.

Ultimate design thrust N_{Ed} = 214.9 kN.

Initial sizing of section

Try a section UB610 × 305 × 179 kg/m; grade of steel S 275; $m_{fy} = 275$ N/mm^2. The properties of the section are as follows.

Depth of section $D = 620.2$ m.

Depth between fillets $d = 540$ mm.

Width of flange $b = 307.1$ mm.

Thickness of web $t = 14.1$ mm.

Thickness of flange $T = 23.6$ mm.

Root radius $r = 16.1$ mm.

Radius of gyration, x–x axis, $i_y = 25.9$ cm.

Radius of gyration, y–y axis, $i_z = 7.07$ cm.

Elastic modulus, x–x axis, $Z_x = 4940$ cm^3.

Elastic modulus, y–y axis, $Z_y = 743$ cm^3.

Plastic modulus, x–x axis, $S_x = 5550$ cm^3.

Plastic modulus, y–y axis, $S_y = 1140$ cm^3.

Area of section $A = 228$ cm^2.

$b = B/2 = 307.1/2 = 153.6$.

$b/T = 153.6/23.6 = 6.51$, $d/t = 38.3$ and $D/T = 620.2/23.6 = 26.3$.

Section classification

Firstly, before designing the section, we have to classify the section into one of the following classes:

- *Class 1, plastic:* cross-section with hinge rotation capacity.
- *Class 2, compact*: cross-section with plastic moment capacity.
- *Class 3, semi-compact*: cross-section in which the stress in the extreme compression fibre can reach the design strength but a plastic moment capacity cannot be developed.
- *Class 4, slender*: cross-section for which we have to have a special allowance owing to the effects of local bending.

Let us consider our assumed section UB610 × 305 × 79 kg/m; grade of steel S 275. Referring to Table 9 of BS 5950: 2000-1 ("Design strength p_y),

thickness of flange $T = 23.6$ mm

Since the thickness of the flange is greater than 16 mm and less than 40 mm, the design strength $p_y = 265$ N/mm^2. Referring to Table 11 of BS 5950: 2000-1 ("Limiting width-to-thickness ratio for section other than CHS and RHS"),

$$\varepsilon = (275/p_y)^{0.5} = (275/265)^{0.5} = 1.02$$

For class 1, plastic, the limiting value $9\varepsilon \geq b/T$ (6.51). In our case,

$$9 \times 1.02 = 9.2 > 6.51$$

Also, for a web with its central axis at mid depth, for a class 1 plastic section the limiting value of $80\varepsilon \geq d/t$ (38.3). In our case,

$$80 \times 1.02 > 38.3$$

Therefore the section chosen is a class 1 plastic section. Satisfactory

To check shear capacity of section

Referring to Clause 4.2.3 of BS 5950: 2000-1 ("Shear capacity"), the shear force F_v should not be greater than the shear capacity P_v of the section. In our case,

$$P_v = 0.6p_yA_v$$

where A_v = shear area = $tD = 14.1 \times 620.2 = 8745$ mm^2, and so

$$P_v = 0.6 \times 265 \times 8745/10^3 = 1390 \text{ kN} > F_v \text{ (67.8 kN)}$$

Since d/t (38.3) $< 70\varepsilon$ (71.4), the section need not be checked for shear buckling.

To check against shear buckling

Since d/t (38.3) $< 70\varepsilon$ (71.4), the section need not be checked for shear buckling.

To check moment capacity of section

Referring to Clause 4.2.5.2 of BS 5950: 2000-1 ("Moment capacity with low shear"), if the shear force F_v does not exceed 60% of the shear capacity P_v, then, for class 1 and class 2 sections, the moment capacity $M_c = p_yS_x$. In our case, $F_v = 67.8$ kN and 60% of $P_v = 0.6 \times 1390 = 834$ kN. So, $F_v < 0.6P_v$. Therefore

$$M_c = p_yS_x < 1.2p_yZ_x = 265 \times 5550/10^3 < 1.2 \times 265 \times 4940/10^3$$
$$= 1471 < 1571$$

We adopt the lower value.

Therefore the ultimate design moment M_u (895.2 kN m) $< M_c$ (1471). Satisfactory

Actual bending stress $f_{bc} = 895.2 \times 10^6/(5550 \times 10^3) = 161$ N/mm^2.

To check for lateral–torsional buckling

Referring to Clause 4.3 of BS 5950: 2000-1 ("Lateral–torsional buckling"), as already discussed, the effective length of the unstrained bottom compression flange Ly is 3.0 m. Referring to Clause 4.3.6.7 ("Equivalent slenderness λ_{LT}"),

$$\text{effective slenderness } \lambda_{LT} = \lambda_y uv\beta_w{}^{0.5}$$

where $\lambda_y = y$–y axis slenderness $= L_y/r_y = 300/7.07 = 42.4$, and u is the buckling parameter; referring to Clause 4.3.6.8 ("Buckling parameter u and torsional index x), for rolled I and H sections or channels with equal flanges, and with $x = D/T = 20.4$, $u = 0.9$. v is the slenderness factor for the beam. With $D/T = 20.4$ and $\lambda_y/x = 33/20.4 = 1.6$, referring to

Table 19, $v = 0.97$. β_w is a ratio; referring to Clause 4.3.6.9 for class 1 and class 2 sections, $\beta_w = 1.0$. Therefore

effective slenderness $\lambda_{LT} = \lambda_y uv\beta_w^{0.5} = 42.4 \times 0.9 \times 0.97 \times 1.0 = 37$

Referring to Table 16 of BS 5950: 2000-1 ("Bending strength p_b for rolled section"), and with $\lambda_{LT} = 37$ and $p_y = 265$ N/mm^2,

allowed bending stress $p_b = 254$ N/mm^2.

Therefore

buckling moment of resistance $M_b = p_b S_x = 254 \times 5550/10^3 = 1410$ kN m $> M_u$ (895.2 kN m).

Ratio = 895.2/1410 = 0.63.

In addition, the column is subjected to compression.

To check for compression

Ultimate design compression $F_c = 214.9$ kN.

On the minor axis, the side rails run at 2.0 m centres. So, the slenderness λ_y for the minor axis is 200/7.07 = 28. Referring to Table 24 of BS 5950: 2000-1 for p_c in N/mm^2 with $\lambda \leq 110$, with $\lambda_y = 28$, the compressive strength p_c is 258 N/mm^2.

Actual compressive stress $f_c = F_c/_A = 214.9 \times 10^3/(228 \times 10^2) = 9.4$ N/mm^2.

Therefore

$f_c/p_c + f_{bc}/p_b = 9.4/258 + 161/254 = 0.04 + 0.63 = 0.67 < 1.0$.

In conclusion, when the results obtained following Eurocode 3 and BS 5950 are compared, there is hardly any difference.

References

British Standards Institution, 1983. BS 2573-1: 1983, Rules for the design of cranes. Specification for classification, stress calculations and design criteria for structures.

British Standards Institution, 2000. BS 5950-1: 2000, Structural use of steelwork in building. Code of practice for design. Rolled and welded sections.

British Standards Institution, 2004. BS EN 10025-1: 2004, Hot rolled products of structural steels. General technical delivery conditions.

Eurocode, 2002a. BS EN 1990: 2002(E), Basis of structural design.

Eurocode, 2002b. BS EN 1991-1-1: 2002, Actions on structures. General actions. Densities, self-weight, imposed loads for buildings.

Eurocode, 2003. BS EN 1991-1-3: 2003, Actions on structures. General actions. Snow loads.

Eurocode, 2005a. BS EN 1993-1-1: 2005, Eurocode 3. Design of steel structures. General rules and rules for buildings.

Eurocode, 2005b. BS EN 1991-1-4: 2005, Actions on structures. General actions. Wind actions.
Eurocode, 2005c. BS EN 1993-1-8: 2005, Eurocode 3. General. Design of joints.
Eurocode, 2006. BS EN 1991-3: 2006, Actions on structures. Actions induced by cranes and machines.
Stanton, T.E., 1908. Experiments on wind pressure. *Minutes of the Proceedings of the Institution of Civil Engineers*, 171, 175–200.

CHAPTER 3

Design of Gantry Girders (Members Subjected to Biaxial Bending)

In this chapter, we examine the design of a welded-plate gantry girder in the melting bay.

3.1 Design philosophy

The term "gantry girder" may be defined as meaning a structural beam section, with or without an additional plate or channel connected to the top flange to carry overhead electric travelling cranes. Normally, for medium-duty (say 25 to 30 t capacity) cranes, standard universal rolled I-beams are used.

In the design of gantry girders with long spans supporting heavy vertical dynamic crane wheel loads with transverse horizontal crane surges, the standard universal rolled beam section is not adequate as a gantry girder; the built-up section of a plate girder is adopted instead. Previously, plate girders were generally made up of web plates and flanges consisting of angles and plates, and fabricated by use of riveted connections; fabrication was time-consuming. With the advent of modern automated welding technology in recent years, however, almost all built-up plate girders, composed of web and flange plates of any designed dimensions, can be shop fabricated by welding.

In our case, the gantry girders under consideration support an overhead electric travelling crane of 2900 kN/600 kN capacity that carries a molten-steel pot. The girders are subjected to very high vertical impacts, transverse horizontal surges and longitudinal horizontal surges. This is due to the sudden application of brakes of a heavily loaded moving crane, possible slip of a sling with a load, cross-travelling of crabs with a load, and also the dragging of loads across the working floor. Normal rolled universal beams will not be adequate to resist moments and shear for supporting this type of travelling crane.

Therefore, we shall consider a *built-up welded-plate girder*. The girder may be simply supported or continuous over supports. Continuity over supports reduces the depth and cost, but any differential settlement of the supports may reverse the original design values of the moments in the sections, thus exceeding the allowable stress in the material, with consequent collapse of the member. So, we adopt simply supported gantry girders, particularly for very heavily loaded girders.

In the design considerations, theoretically most of the shear force on the section is taken by the web plate. It may therefore be assumed in the plate girder design that the top and bottom flanges resist the whole moment and that the web takes the whole shear at the section considered.

3.2 Detailed considerations

3.2.1 Effective span of girder

The effective span is equal to the centre-to-centre distance between the bearings. The spacing of the stanchions varies from 18 m to 24 m. The variation of the column spacing is due to the mechanical layout of machines, and other requirements. For design purposes, we assume the longest span $L = 24$ m (centre-to-centre distance of stanchions), simply supported.

$$\text{Effective span} = \text{centre-to-centre distance of bearings}$$
$$= L_e = 24.0 - 2 \times 0.310 = 23.38$$

say 23.4 m (where the supporting column is assumed to be UB914 × 305 × 289 kg/m (a stanchion composed of 2 × UB914 × 305 × 289 kg/m), and the distance from the centre line of the column to the centre of the bearing is 310 mm).

3.2.2 Gantry girder loading design data

Gantry girder design data is generally supplied by crane manufacturers. In the absence of information from the manufacturers, the data may be obtained from the British Standard specifications in BS 466: 1960 and BS 2573-1: 1966 and subsequent more recent revised codes (see British Standards Institution, 1983, 1984).

The crane data supplied by the manufacturer is given below:

- Crane capacity = 2900 kN/60 kN/10 kN.
- Maximum load lifted = 2900 kN.
- Crane span = span of truss − (2.5 m + 1.5 m assumed) = 28.50 − 4.0 (assumed) = 24.5 m.
- Maximum wheel load = 520 kN.
- Number of wheels in crossheads or end carriage = 4.
- Number of end carriages on each end of crane = 2 spaced at 10.2 m.
- Dead weight of crane = 2700 kN (shared equally by 16 wheels, 8 wheels in each end carriage).
- Therefore minimum wheel load = 2700/16 = 168.8, say 170 kN.
- Spacing of wheels = 1.2 m, 1.5 m, 1.2 m as shown in Fig. 3.1.
- End clearance of crane = 600 mm (minimum).
- Minimum headroom from rail top = 4500 mm.
- Weight of crab = (1/5 of maximum load lifted + 5 kN) = (1/5) × 2900 + 5 = 585 kN.

3.2.3 Vertical dynamic impact factor

The dynamic impact factor for vertical static wheel loads depends on the type of crane, class and duty factor. For overhead electric travelling cranes of class 3 (heavy foundry work with a duty factor of 0.9), BS 2573 specifies an impact factor of 1.4. In Eurocode 1, Part 3 (Eurocode, 2006), for hoisting class HC4 (heavy-duty cranes in foundries, with intermittent grab, magnet grab and forging), the dynamic factor φ is also normally 1.4.

In our case, the crane manufacturer provided a vertical dynamic impact factor $\varphi = 1.4$. So we adopt a vertical impact factor for wheel loads of 1.4. Therefore we adopt

$$\text{maximum vertical dynamic wheel load factor } W_{\text{v dyna}} = 1.4 \times 520 = 728 \text{ kN}.$$

3.2.4 Transverse horizontal surge

Eurocode 1, Part 3, specifies:

$$\text{transverse horizontal surge} = (1/10) \times (\text{crane capacity} + \text{weight of crab})$$
$$= (1/10) \times (2900 + 585) = 348.5 \text{ kN}.$$

The whole horizontal surge should be equally distributed across all 16 wheels (8 wheels on each carriage). Therefore

$$\text{transverse horizontal surge on each wheel} = H_{\text{ts}} = 348.5/16 = 21.78, \text{ say } 22 \text{ kN}.$$

3.2.5 Longitudinal tractive force

The longitudinal tractive force is generally taken equal to 1/20 of the maximum wheel load. Therefore the maximum longitudinal tractive force on each wheel = $H_{\text{ls}} = 1/20 \times 520 = 26$ kN. So

$$\text{total longitudinal tractive force on each girder} = 26 \times 8 = 208 \text{ kN}.$$

3.2.6 Moment influence lines

The maximum moment in a section of a girder occurs when the centre of the span divides the distance between the centre of gravity of the load system and the nearest load under consideration. The centre of gravity of the load system is between the nearest wheels of the two end carriages at a distance of 6.3 m. So the centre of the span will divide the distance between the centre of gravity of the load system and the nearest load under consideration equally, at a distance of 6.3/2 = 3.2 m (see Fig. 3.1(a)) (Salmon, 1948). Therefore the load system will be placed with the nearest load of 520 kN at a distance of 1.6 m from the centre of the span.

Effective span = 23.4 m, as calculated before.

Half of span = 11.7 m.

The load under consideration will be placed at a distance of (11.7 + 1.6) m = 13.3 m from the right support, and (11.7 − 1.6) = 10.1 m from the left support. Therefore

maximum ordinate of the moment influence line

$$m_1 = 13.3 \times 10.1/23.4 = 5.74,$$

$$m_2 = 5.74 \times (10.1 - 1.2)/10.1 = 5.05,$$

$$m_3 = 5.74 \times (10.1 - 1.2 - 1.5)/10.1 = 4.2,$$

$$m_4 = 5.74 \times (10.1 - 1.2 - 1.5 - 1.2)/10.1 = 3.52,$$

$$m_5 = 5.74 \times (6.3)/13.3 = 2.72,$$

$m_6 = 5.74 \times (6.3 - 1.2)/13.3 = 2.2,$

$m_7 = 5.74 \times (6.3 - 1.2 - 1.5)/13.3 = 1.55,$

$m_8 = 5.74 \times (6.3 - 1.2 - 1.5 - 1.2)/13.3 = 1.04.$

Therefore

$$\sum m = (5.74 + 5.05 + 4.2 + 3.52 + 2.72 + 2.2 + 1.55 + 1.04) = 26.02.$$

3.2.7 Shear influence lines

The maximum ordinate of the shear influence line will occur at the support when the first wheel of the end carriage is at the support (see Fig. 3.1). Therefore maximum shear ordinate at support

$s_1 = 1,$

$s_{12} = (23.4 - 1.2)/23.4 = 0.95,$

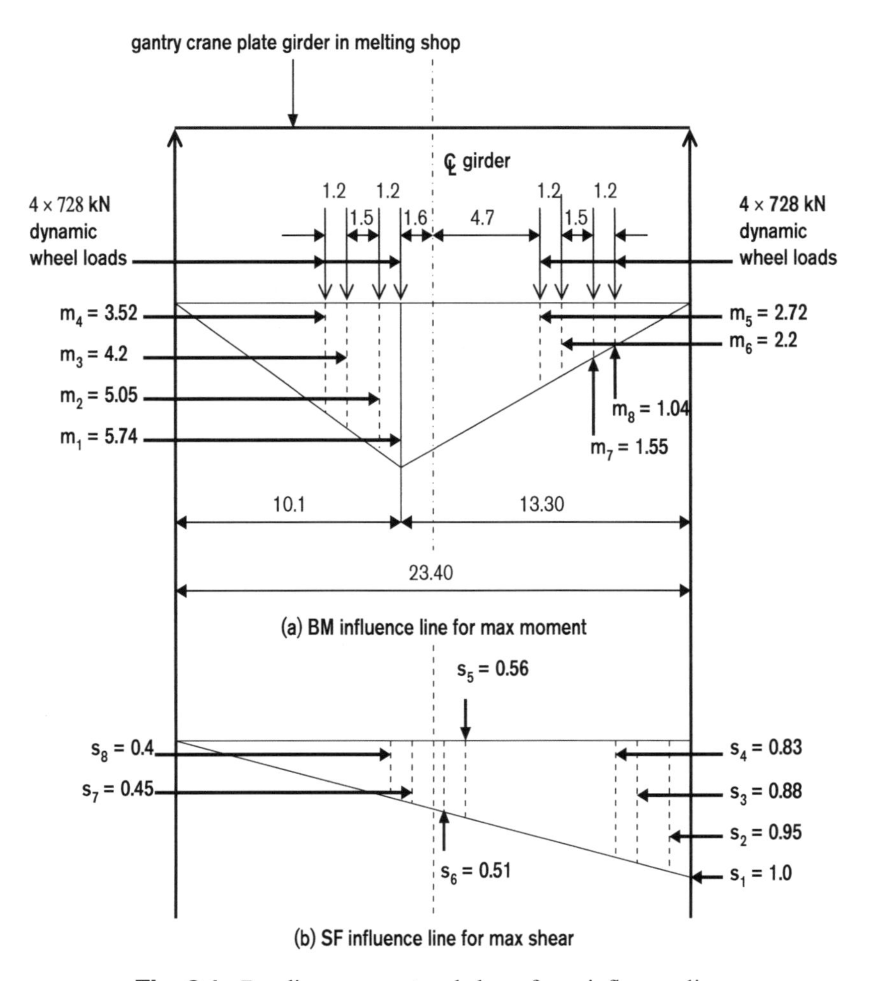

Fig. 3.1. Bending-moment and shear-force influence lines

$s_{13} = (23.4 - 1.2 - 1.5)/23.4 = 0.88,$

$s_{14} = (23.4 - 1.2 - 1.5 - 1.2)/23.4 = 0.83,$

$s_{15} = (23.4 - 3.9 - 6.3)/23.4 = 0.56,$

$s_{16} = (23.4 - 3.9 - 6.3 - 1.2)/23.4 = 0.51,$

$s_{17} = (23.4 - 3.9 - 6.3 - 1.2 - 1.5)/23.4 = 0.45,$

$s_{18} = (23.4 - 3.9 - 6.3 - 1.2 - 1.5 - 1.2)/23.4 = 0.4.$

Therefore

$$\sum s = (1 + 0.95 + 0.88 + 0.83 + 0.56 + 0.51 + 0.45 + 0.4) = 5.58.$$

3.2.8 Characteristic maximum dynamic vertical moment

The maximum dynamic vertical moment in the section 10.1 m from the left support is equal to the sum of all moment influence line ordinates multiplied by the dynamic wheel load. Therefore

$$\text{maximum moment at 10.1 m from left support} = M_{\text{v dynm}}$$
$$= W_{\text{v dynm}} \sum m = 728 \times (5.74 + 5.05 + 4.2 + 3.52 + 2.72 + 2.2 + 1.55 + 1.04)$$
$$= 18\,942 \text{ kN m}.$$

3.2.9 Characteristic maximum dynamic vertical shear at support

The maximum dynamic vertical shear at the support is equal to the sum of all vertical influence line ordinates multiplied by the dynamic wheel load. Therefore

$$\text{maximum dynamic shear at support} = V_{\text{v dynm}} \sum s$$
$$= 728 \times (1.0 + 0.95 + 0.88 + 0.83 + 0.56 + 0.51 + 0.45 + 0.4)$$
$$= 728 \times 5.58 = 4062 \text{ kN}.$$

3.2.10 Characteristic minimum vertical shear

This is due to the minimum wheel load. From shear influence lines,

$V_{\text{min}} = 520$ kN.

Minimum shear at support = $170 \times 5.58 + (\text{crab weight})/8 \times 5.58$

$= 948.6 + 585/8 \times 5.58 = 1357$ kN.

3.2.11 Characteristic vertical design moment due to self-weight of gantry girder

To calculate the moment, we have to evaluate the approximate weight of the girder. The depth of the girder is generally assumed to be between 1/10 and 1/12 of the span.

Effective span of girder $L_{\text{e}} = 23.4$ m.

Assumed depth of girder = $(1/10) \times 23.4 = D = 2.34$ m.

The girder is assumed to be a built-up plate girder, as such a depth of a normal universal beam (UB) is not available. Now,

ultimate vertical design moment = M_{vu} = 28 413 kN m as calculated above.

Approximate force in the compression flange = M_{vu}/D = 28 413 × 10^6/2340
= 12 142 308 N.

Using S 275 grade steel and assuming a 50 mm (>40 mm) flange plate,

design strength = f_y = 255 N/mm^2.

Therefore

area of steel required = A_g = 12 142 308/255 = 47 617 mm^2.

Assuming a 50 mm flange plate,

width of flange plate required = 47 617/50 = 952 mm.

Assume the thickness of the web plate is 20 mm. Therefore

self-weight of girder = weight of 2 flange plates + weight of web

= 2 × 24 × 900 × 50 @ 353 kg/m + 1 × 24 × 2240 × 20 @ 352kg/m = 25 392 kg, say 254 kN.

Add 10% for (stiffeners + crane rail).

Total characteristic dead weight of girder = 254 × 1.1 = 279 kN.

Characteristic dead load moment = M_{du} = $WL_e/8$ = 279 × 23.4/8 = 816 kN m.

3.2.12 Characteristic vertical dead load shear

Characteristic shear due to self-weight of girder = 279/2 = 140 kN.

3.2.13 Total ultimate vertical design moment (ULS method)

Referring to Table A1.2(B) ("Design values of actions (STR/GEO) (Set B)") of BS EN 1990: 2002 (Eurocode, 2002) (see Appendix B), in the ultimate-limit-state (ULS) design method, partial factors are to be used in load combinations as follows:

$$\gamma_{Gj} G_{kj} + \gamma_{Q,1} Q_{k,1}$$

where γ_{Gj} is the partial factor for the permanent load (unfavourable conditions), recommended value = 1.35. G_{kj} is the permanent load (dead load). $\gamma_{Q,1}$ is the partial factor for the leading variable load (an imposed, moving load), recommended value = 1.5. $Q_{k,1}$ is the leading variable load.

Total ultimate vertical design moment

= 1.5 × characteristic moment due to wheel loads
+ 1.35 × characteristic moment due to self-weight:

M_{vu} = 1.5 × 18 942 + 1.35 × 1.35 × 816 = 29 515 kN m.

3.2.14 Total ultimate vertical design shear (ULS method)

Total ultimate vertical design shear

= 1.5 × characteristic shear due to wheel load

+ 1.35 × characteristic shear due to self-weight of girder:

$V_{vu} = 1.5 \times 4062 + 1.35 \times 140 = 6282$ kN.

3.2.15 Maximum ultimate horizontal transverse moment

Referring to Tables A1.2(B) and Table A1.1 of BS EN 1990: 2002(E) for the partial factors acting in conjunction with vertical loads,

partial factor $= \gamma_{Q,1}\psi_{0,1} = 1.5 \times 0.7 = 1.05$.

Therefore

$M_{hu} = 1.05 \times$ (horizontal transverse load) × (∑ moment influence line ordinates)

$= 1.05 \times 22 \times 26.02 = 601$ kN m.

3.2.16 Maximum ultimate horizontal longitudinal tractive force

$F_{hu} = 1.5 \times 208 = 312$ kN.

3.3 Design of section

The section will be designed as a welded-plate girder. Eurocode 3, Part 1-1 (Eurocode, 2005) will be followed. The tables and figures referred to below can be found in Annex A of the Eurocode (Appendix B of this book), except where otherwise mentioned.

3.3.1 Design strength

By referring to Table 3.1 of Eurocode 3 ("Nominal values of yield strength f_y and ultimate tensile strength f_u for hot rolled structural steel"), the design strength (f_y) in the ULS method of design for the flanges and web can be obtained; its value varies with the thickness of plate considered.

In our case, we adopt steel grade S 275 with a design yield strength $f_y = 275$ N/mm^2. So, for a nominal plate thickness $t \leq 40$ mm, $f_y = 275$ N/mm^2, and for $t \leq 80$ mm, $f_y = 255$ N/mm^2. If the design strength of the web f_{yw} is greater than the design strength of the flange f_{yf}, then the design strength of the flange should always be used when considering moments or shear.

3.3.2 Initial sizing of section

The dimensions of the webs and flanges are assumed to be as given below.

Overall depth of girder = h. The depth should be chosen to limit the allowable deflection. In practice, the overall depth should normally be taken to be between 1/10 and 1/12 of the span. In our case, we assume the overall depth h = 1/10 of span = 23.4/10 = 2.34 m = 2340 mm. We assume an overall depth h = 2500 mm (as the girder is subjected to high dynamic wheel loads).

Depth of straight portion of web $d = h - 2 \times$ size of weld $- 2 \times$ thickness of flange $= 2500 - 2 \times 12$ (assumed weld size) $- 2 \times 55$ (assumed) $= 2366$ mm.

Breadth of flange $= b$. The breadth of the flange should be at least 1/40 to 1/30 of the span in order to prevent excessive lateral deflection. In our case, we assume a breadth $b = 1/30$ of span $= (1/30) \times 23.4 = 0.78$ m, say 0.9 m $= 900$ mm.

Thickness of web $= t_w$. Several tests have shown that the web does not buckle owing to diagonal compression when the ratio d/t_w is less than 70, if the web is not stiffened by a vertical transverse stiffener. Referring to Table 5.2 (sheet 1) of Eurocode 3, Part 1-1, the minimum thickness of web required to avoid buckling of the compression flange in the ULS design method with an unstiffened web is as follows. For class 1 classification, $d/t_w \leq 72\varepsilon$, where

$$\varepsilon = \text{stress factor} = (235/f_y)^{0.5} = (235/255)^{0.5} = 0.96;$$

$$d/t_w = 2366/t_w = 72 \times 0.96.$$

Therefore

$$t_w = 2366/(72 \times 0.96) = 34 \text{ mm}.$$

With a stiffened web and a spacing of transverse stiffeners $a \leq d$,

$$t_w \geq (d/250)(a/d)^{0.5}.$$

Assuming a spacing of stiffeners $a = 2366$ mm,

$$t_w = 2366/250 \times (2366/2366)^{0.5} = 10 \text{ mm}.$$

We assume $t_w = 30$ mm.

Thickness of flange $= t_f$. The minimum thickness of the flange required to limit the outstand of the flange is calculated as follows. The approximate flange area required is given by

$$A_f = M_{vu}/(hf_y) = 295\ 158 \times 10^6/(2500 \times 255) = 46\ 298 \text{ mm}^2.$$

Assuming the width of the flange $b = 900$ mm,

$$t_f = 46\ 298/900 = 51.4 \text{ mm}.$$

We therefore assume $t_f = 55$ mm.

3.3.3 Classification of cross-sections

Referring to Clause 5.5.2 of Eurocode 3, Part 1-1, the function of cross-section classification is to identify the extent to which the resistance and rotation capacity of the cross-section are limited by its local buckling resistance. Cross-sections are classified into four categories as described below:

- *Class 1:* the cross-sections in this class are those which can form a plastic hinge with the rotation capacity required from plastic analysis without reduction of resistance.

- *Class 2:* the cross-sections in this class are those which can develop plastic moment resistance but have limited rotation capacity because of local buckling.
- *Class 3:* the cross-sections in this class are those in which the stress in the extreme compression fibre of the steel member, assuming an elastic distribution of stress, can reach the yield point, but local buckling prevents the development of plastic moment resistance.
- *Class 4:* the cross-sections in this class are those in which local buckling will occur before the full yield stress is reached in one or more parts of the cross-section.

In our case, we assume a class 1 cross-section classification without reduction of resistance. Thus, to determine the thickness t_f of the flange, we do the following.

For class 1 section classification, $c/t_f \leq 9\varepsilon$, where

$$c = \text{outstand of flange plate} = [b - (t_w + 2 \times 12 \text{ (weld size)})]/2 = [900 - (25 + 24)]/2 = 425.5 \text{ mm.}$$

Assuming $t_f = 55$ m,

$$c/t_f = 425.5/55 = 7.7 \text{ and } 9\varepsilon = 9 \times 0.96 = 8.64$$

Since c/t_f (7.7) < 9ε (8.64), the section satisfies the conditions for class 1 section classification. So we assume $t_f = 55$ mm.

To determine the thickness of the web t_w, we do the following.

For class 1 section classification, $d/t_w \leq 72\varepsilon$. Assuming $t_w = 30$ mm,

$$d/t_w = 2366/30 = 78.9 \text{ and } 72\varepsilon = 72 \times 0.96 = 69 < d/t_w$$

which does not satisfy the condition. We increase the thickness t_w to 35 mm:

$$d/t_w = 2366/35 = 67.6 < 72\varepsilon\,(69)$$

which satisfies the condition. So, we assume $t_w = 35$ mm.

Thus the initial sizing of the section is as follows:

- Depth of girder $h = 2500$ mm.
- Breadth of flange $b = 900$ mm.
- Depth of straight portion of web $d = 2500 - 2 \times 55 - 2 \times 12$ (weld size) $= 2366$ mm.
- Thickness of web $t_w = 35$ mm.
- Thickness of flange $t_f = 55$ mm.
- Design strength with flange thickness 55 mm $= f_y = 255$ N/mm^2.
- Design strength of web with thickness 35 mm $= f_y = 255$ N/mm^2 (see below).

Although the design strength of the web is 275 N/mm^2 for 35 mm thickness, the lower value of f_y (255 N/mm^2) of the flange should be considered in calculations for moments and shears (see Fig. 3.2).

3.3.4 Moment capacity

Total maximum ultimate vertical design moment $= M_{vu} = 29\,515$ kN m.

Total maximum design shear $= V_u = 6281$ kN.

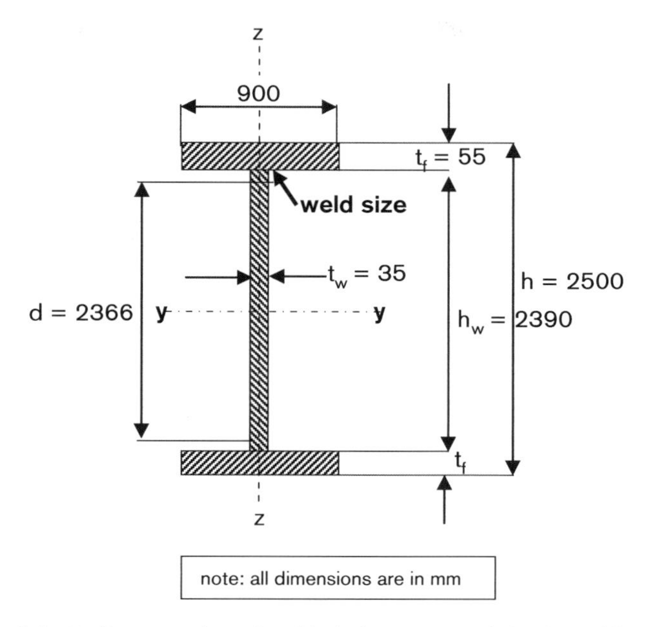

Fig. 3.2. Built-up section of welded-plate gantry girder in melting bay

The moment capacity should be calculated in the following way.

When the web depth-to-thickness ratio $d/t_w \leq 72\varepsilon$, it should be assumed that the web is not susceptible to buckling, and the moment capacity should be calculated from the equation

$$M_{rd} = f_y W_{pl}$$

provided the shear force V_{Ed} (V_{vu}) $\leq 0.5V_{pl,rd}$ (shear capacity), where M_{rd} = moment capacity, W_{pl} = plastic section modulus and $V_{pl,rd}$ = shear capacity. In our case, d/t_w (67.6) $< 72\varepsilon$ (69.1).

Thus, the web is not susceptible to buckling.

The ultimate shear force (V_{vu}) should also be less than half the shear capacity ($V_{pl,rd}$) of the section. Referring to equation (6.18) of Eurocode 3, Part 1-1,

$$V_{pl,rd} = A_v[f_y/(3)^{0.5}]/\gamma_{Mo}$$

where A_v = shear area

$$= dt_w + (t_w + 2r)t_f$$

$$= 2366 \times 25 + (25 + 2 \times 12) \times 50 = 61\,600 \text{ mm}^2.$$

Referring to Clause 6.1, γ_{Mo} = partial factor = 1.0, and f_y = 335 N/mm^2 (because $t_f > 40$ mm). Therefore

plastic shear capacity $V_{pl,rd} = 61\,600 \times [335/(3)^{0.5}]/1.0/10^3 = 12\,669$ kN

and $0.5V_{pl,rd} = 12\,669/2 = 6335$ kN $> V_{vu}$ (6281 kN).

Thus, V_{vu} (6281 kN) $< 0.5V_{pl,Rd}$ (6335 kN).

So the section satisfies the conditions.

Since the web is not susceptible to buckling, and the lowest shear value in the section is less than half the shear capacity of the section, the moment capacity for this class 1 compact section should be determined by the "flange only" method. In this case, the whole moment will be taken up by the flanges alone and the web takes the shear only. Therefore

moment capacity of section $M_{y,\mathrm{Rd}} = f_y A_f h_s$

where A_f = area of compression flange = $b \times t_f = 900 \times 55 = 49\,500\ \mathrm{mm}^2$, h_s = depth between centroids of flanges = 2500 − 55 = 2445 mm and $f_y = 255\ \mathrm{N/mm}^2$. Therefore

$$M_{y,\mathrm{Rd}} = 255 \times 49\,500 \times 2445/10^6 = 30\,862\ \mathrm{kN\ m} > M_{\mathrm{Ed}}\ (M_{\mathrm{vu}})\ (29\,515\ \mathrm{kN\ m}).$$

Satisfactory

Alternatively, referring to Clause 6.2.5, the moment capacity of the section may be expressed by the following equation:

$$M_{\mathrm{pl,Rd}} = W_{\mathrm{pl}} f_y/\gamma_{\mathrm{Mo}}, \qquad (6.12)$$

where W_{pl} is the plastic modulus of the section. (The equation numbers in this chapter refer to Eurocode 3, Part 1-1.) As the section assumed is built-up welded and of high depth, no rolled I section is available of this depth. The plastic modulus for the assumed depth is not easy to calculate. So the above equation can only be used when the assumed section is manufactured industrially.

In addition, the top flange is also subjected to a stress due to the horizontal transverse moment caused by horizontal crane surges. Therefore, the "flange only" method is suitable in our case. The horizontal transverse ultimate moment M_{hu} is equal to 601 kN m (calculated previously). This transverse horizontal moment is resisted by a horizontal girder formed by the connection of the 6 mm plate (acting as a web) of the walking platform (Durbar), between the top flange of the main plate girder and the tie beam at the walking-platform level.

Distance between centre line of plate girder and the tie beam = h_z = 2.5 m

Horizontal moment of resistance = $M_{z,\mathrm{Rd}} = f_y A_f h_z$

where A_f = area of top flange = $900 \times 55 = 49\,500\ \mathrm{mm}^2$.

Therefore

$$M_{z,\mathrm{Rd}} = 255 \times 49\,500 \times 2500/10^6 = 315\,56\ \mathrm{kN\ m} \gg M_{\mathrm{hu}}\ (601\ \mathrm{kN\ m}).$$

Referring to the criterion based on the quantity

$$[M_{y,\mathrm{Ed}}/M_{y,\mathrm{Rd}}]^{\alpha} + [M_{z,\mathrm{Ed}}/M_{z,\mathrm{Rd}}]^{\beta} \qquad (6.41)$$

where α and β are constants, which may conservatively be taken as unity, this quantity is equal to

$[29\,515/30\,862] + [601/31\,556] = 0.95 + 0.02 = 0.97 < 1$ Satisfactory

Therefore we adopt the following section for the welded-plate girder:

- Depth h = 2500 mm.
- Breadth b = 900 mm.

- Thickness of flange $t_f = 55$ mm.
- Thickness of web $t_w = 35$ mm.

3.3.5 Moment buckling resistance

When a structural member is subjected to bending due to external load actions, the unrestrained length of the compression flange moves laterally at right angles to its normal bending and rotational displacement, thus creating a lateral–torsional buckling of the member. For example, consider a simply supported beam, loaded as shown in Fig. 3.3. The ends of the beam are restrained laterally but the compression top flange is not laterally restrained for the whole span. As a result, when vertical bending occurs with a vertical deflection Δ_v, the compression flange deflects laterally by Δ_h and, at the same time, a rotational displacement α° takes place as shown in the figure. Thus, not only does a vertical bending moment occur in the member owing to external loadings, but the member also undergoes a torsional buckling moment. So we have to investigate if the member is capable of resisting the torsional buckling moment in addition to the vertical bending moment. The laterally unrestrained compression flange, when subjected to major-axis bending, should be verified against lateral–torsional buckling.

In our case, the top compression flange is restrained by the chequered plate of the walkway, which is securely fixed to the top flange. So, the chequered plate, acting with stiffeners at intervals of 2.4 m, restrains the buckling of the top flange (see Fig. 3.10). Therefore, there is no need to verify the buckling resistance of the top flange in bending.

3.3.6 Shear buckling resistance

The shear buckling resistance should be checked if the ratio $d/t_w > 72\varepsilon$ for a plate girder of welded section of class 1 classification. In our case,

$$d/t_w = 2366/35 = 67.3$$

and $72\varepsilon = 72 \times 0.96 = 69.1$, i.e. $d/t_w < 72\varepsilon$

Therefore, the shear buckling resistance need not be checked.

3.3.7 End anchorage

In a plate girder, the web is used as a thin membrane to carry the shear and may buckle owing to loadings when the ratio of web depth to thickness exceeds a limiting value without transverse stiffeners, as described above. The shear capacity is very much limited, so that the shear buckling resistance reaches its maximum allowable value without failure. By bringing transverse stiffeners into place on the web plate, one can increase the design *plastic shear resistance* $V_{pl,rd}$, and thus increase the *buckling shear resistance* $V_{b,Rd}$.

In addition, stiffeners enhances the load-bearing capacity of a plate girder to a large extent by the development of *tension field action*. When the buckling shear resistance of a web panel of a plate girder exceeds the allowable value and the web buckles to the brink of failure, its capacity to carry additional diagonal compressive stress becomes negligible and should be ignored. When stiffeners are used, a new load-bearing mechanism is created within the web, in which we may imagine that an N-truss in a certain zone of the web acts as a diagonal tension member to carry any additional shear, and the vertical stiffeners are in compression. The imaginary tensile force anchors the top and bottom flanges and also anchors the transverse stiffeners on either side of the web panel. This is explained in Section 3.3.8 and Fig. 3.4.

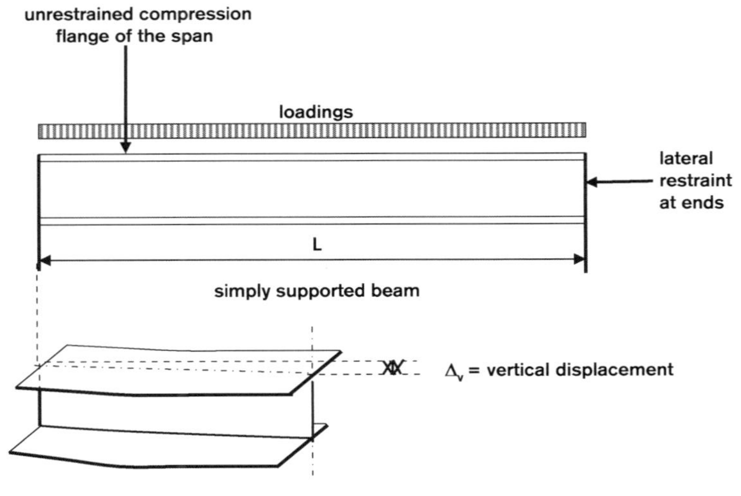

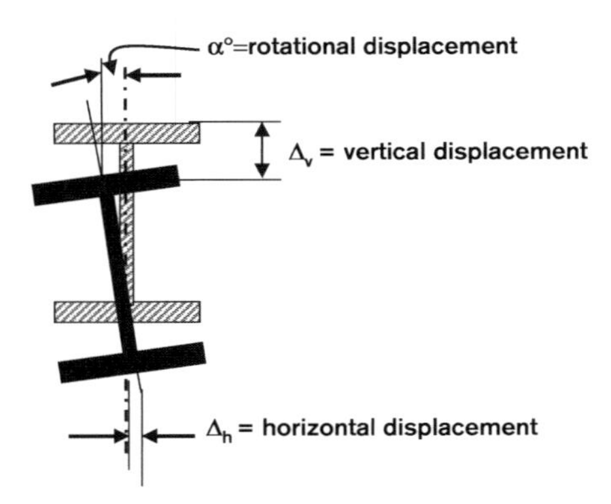

Fig. 3.3. Lateral–torsional buckling of unrestrained compression flange

End anchorage may not be necessary if either of the following conditions is satisfied.

(1) The shear capacity, not the shear buckling resistance, is the governing design criterion, as given by the following condition:

$$V_{b,Rd} = V_{pl,rd}$$

where

$V_{b,Rd}$ = shear buckling resistance

$V_{pl,rd}$ = plastic shear capacity of the member = $A_v(f_y/\sqrt{3})/\gamma_{Mo}$ = 12 669 kN (already calculated),

$$V_{b,Rd} = h_1 t_w q_w$$

where q_w = shear buckling strength of web, which depends on the values of h_1/t_w and s/h_1, and where h_1 = web depth, s = stiffener spacing and t_w = web thickness. For a welded I section,

$$f_v = \text{ultimate shear strength} = f_y/\sqrt{3} = 0.6f_y = 0.6 \times 255 = 153\ \text{N/mm}^2$$

$$\lambda_w = [f_v/q_e]^{0.5}, \text{ where if } s/h_1 \leq 1,\ q_e = [0.75 + 1/(s/h_1)^2][1000/(h_1/t_w)]^2$$

In our case s/h_1 = 2400/2390 = 1.004, which exceeds 1. So, when $s/h_1 > 1$

$$q_e = [1 + 0.75/(s/h_1)^2][1000/(h_1/t_w)]^2$$

$$= [1 + 0.75/(1.004)^2][1000/1.004]^2 = 1.74 \times 992\ 048 = 1\ 726\ 163\ \text{N/mm}^2$$

$$\lambda_w = (f_v/1\ 726\ 163)^{0.5} = 0.01 < 0.8$$

Therefore

$$q_w = f_v = 153\ \text{N/mm}^2$$

Therefore the shear buckling resistance is

$$V_{b,Rd} = h_1 t_w q_w = 2390 \times 35 \times 153\ \text{N}/10^3 = 12\ 798\ \text{kN} > V_{pl,rd}\ (12\ 669\ \text{kN})$$

Thus, we find that the shear buckling resistance is not the governing design criterion.

(2) Sufficient shear buckling resistance is available without forming a tension field, as given by $V_{Ed} \leq V_{cr}$, where V_{Ed} = 6282 kN (ultimate shear at support, already calculated) and V_{cr} = critical shear buckling resistance, as given by the following.

If $V_{pl,rd} = V_{b,Rd}$, then $V_{cr} = V_{pl,rd}$. $V_{pl,rd}$ = 12 669 kN (plastic shear capacity) and $V_{b,Rd}$ = 12 798 kN (shear buckling resistance). If $V_{pl,rd} > V_{b,Rd} > 0.72V_{pl,rd}$, then $V_{cr} = (9V_{b,Rd} - 2V_{pl,Rd})/7$. $V_{pl,rd}$ = 12 669 kN; $V_{b,Rd}$ = 12 798 kN; $0.72V_{pl,rd} = 0.72 \times 12\ 669 = 9122$ kN. So,

$$V_{cr} = \text{critical shear buckling resistance} = (9 \times 12798 - 2 \times 12\ 669)/7 = 12\ 835\ \text{kN}$$

Therefore V_{Ed} (6282) < V_{cr} (12 835). Thus, sufficient critical shear buckling resistance is available without forming a tension field. So, the above condition is satisfied. Therefore, an end anchorage is not necessary.

3.3.8 Web bearing capacity, buckling resistance and stiffener design

We consider the behaviour of the stress distribution in the web of a plate girder. The web of a plate girder is generally made of a thin plate. This thin plate buckles under the action of external direct loads and shear. The behaviour of the stress distribution within the plate is very complex. An exhaustive analysis of the theory of the elastic stability of thin plates has been given by Timoshenko and Gere (1961). We shall use instead a simple method to visualize the behaviour of stresses within a thin web plate.

The web plate is subjected to shear and direct stresses, which results in both diagonal tension and compression acting at right angles to each other within the plate. The compression force tends to buckle the web, which must either be thick enough to resist this buckling or be restrained laterally by means of stiffeners. Let us consider an imaginary lattice girder with alternate diagonal compression and tension members at 45° within the web plate, which is subjected to a shear V in each section. The shear will be resisted by the vertical component of the diagonal compressive force in the web. Thus, the diagonal compressive force is a function of the applied vertical shear. The greater the vertical shear, the greater the diagonal compression. Therefore, if the vertical shear stress in any section of an unstiffened web of a plate girder exceeds the permissible value, there will be a necessity to introduce vertical stiffeners.

For the tension forces, let us imagine an N-truss, which may be supposed to act within the web plate, the diagonals acting as tension members and the stiffeners as compression members (Stewart, 1953) (see Fig. 3.4).

We may compare this with the application of very thin web plates in aeroplane wing construction. The thin skin plate (a membrane) buckles because of elastic instability but does not fail, and still carries the load. The reason is that the shear can now be taken by diagonal tension in the web. We refer the reader to the Wagner diagonal-tension theory.

The stiffener at the end of a girder is intended to carry the total end reaction of the girder as a strut member. The function of the stiffener is not only to support the concentrated load but also to distribute the load from the top flange. The evaluation of the spacing of stiffeners for the elastic stability of such web plates has been studied in recent years, and the limitations on the spacing with respect to the ratio of the depth of the web to the thickness have been established.

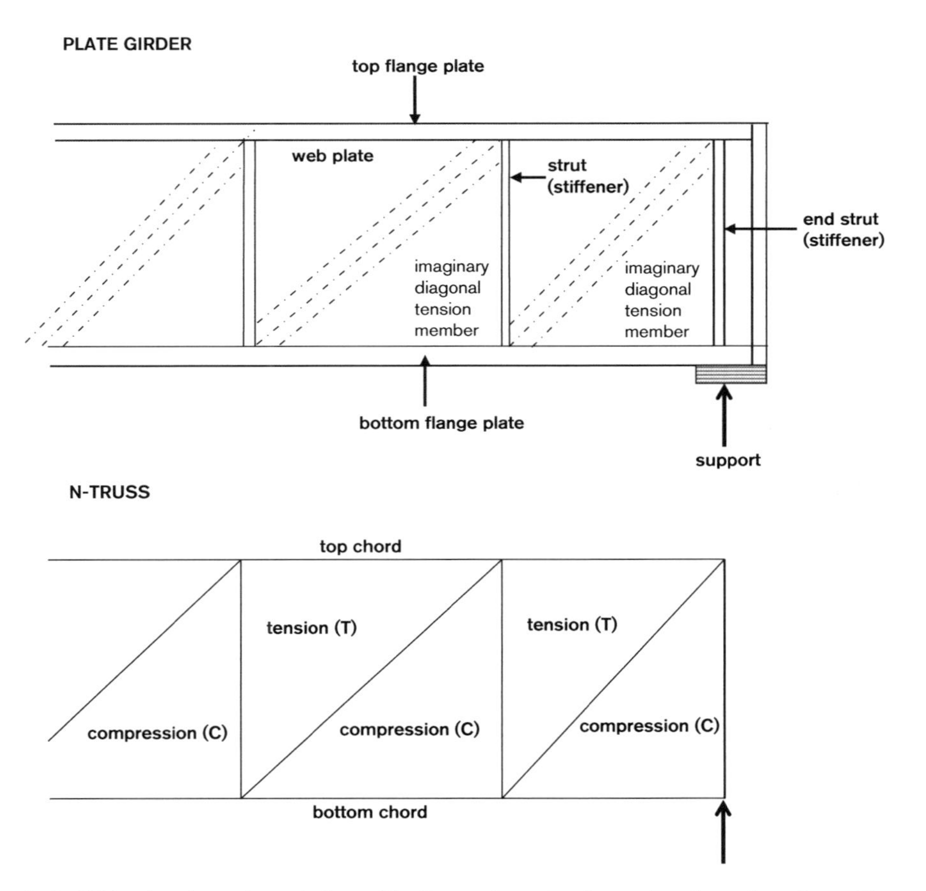

Fig. 3.4. Web plate in a plate girder, with diagonals acting in tension in an imaginary N-truss and the stiffeners acting as struts

3.3.9 Bearing capacity of web

3.3.9.1 Bearing capacity of unstiffened web of gantry girder at the stanchion support

We have to check if the area of the web plate, without stiffeners, at the support of the end of the gantry girder is adequate to transfer the end reaction of the gantry girder at the top flange level due to the ultimate crane wheel loads to the bearing at the bottom flange level (see Fig. 3.5). Bearing stiffeners should be provided where the local compressive force Fx applied through the top flange by the local load of reaction due to the crane wheel loads exceeds the bearing capacity P_{bw} of the unstiffened web at the web-to-flange connection.

In our case, the bottom flange at the end of the girder is connected to the stiff bearing plate of a rocker (or roller) bearing (see Fig. 3.5 again).

The bearing capacity of the unstiffened web may be expressed in the following form:

$$P_{bw} = (b_1 + nk)t_w f_y$$

where b_1 is the assumed width of the stiff bearing plate under the bottom flange = 240 mm, t_w is the thickness of the web = 35 mm and n is a constant, the value which depends on the location. The value is usually taken equal to 5 except at the end of the member.

In our case, the location is at the end of the member. So,

$$n = 2 + (0.6b_e/k) \leq 5$$

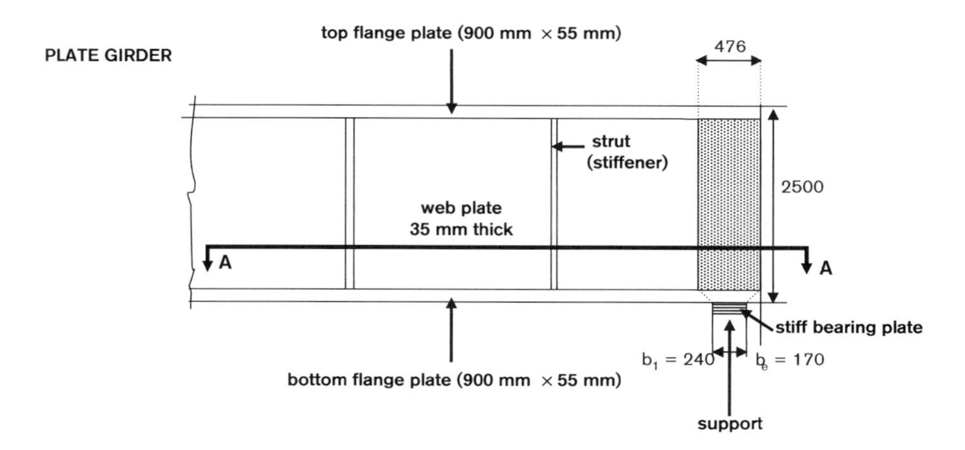

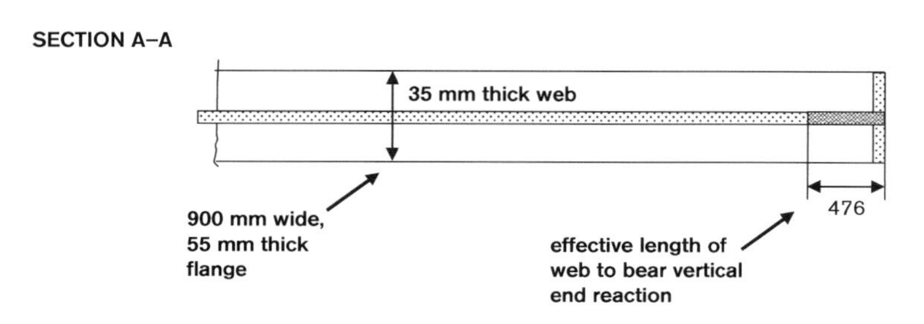

Fig. 3.5. Unstiffened web plate (shaded) of a gantry plate girder to bear end reaction

where

b_e = distance from the nearest end of stiff bearing to the end of girder = 170 mm

$k = t_f$ + size of weld = 55 + 12 (assumed) = 67 mm

$n = 2 + (0.6 \times 120/67) = 3.52 < 5.$ OK

f_y is the assumed yield strength of the web in compression = 255 N/mm^2. Therefore

$$P_{bw} = (240 + 3.52 \times 67) \times 35 \times 255/10^3 = 4247 \text{ kN}$$

and N_{Ed} = compressive force = ultimate end reaction of girder = V_{vu} = 6281 kN (calculated before). So, $P_{bw} < N_{Ed}$. Thus, the bearing capacity of the web without stiffeners is not enough to resist the compressive force due to crane load reactions.

So, we have to introduce end bearing stiffeners.

3.3.9.2 Design of end bearing stiffeners

Consider the behaviour of an end bearing stiffener (see Fig. 3.6). This stiffener takes the whole end reaction of the girder loadings, and acts as a strut member in conjunction with a specified length of web. The effective length of the strut is generally taken to be between 0.7 and 0.75 of the height of the web, assuming the strut to be fixed at the end of the member. The bearing stiffener should be designed for the applied force N_{Ed} (V_{vu}) minus the bearing capacity of the web P_{bw}.

In our case, the bearing capacity of the web is less than the ultimate vertical reaction N_{Ed} (V_{vu}). So, an end bearing stiffener is needed. We assume two end stiffeners made of 30 mm thick plate.

Maximum outstand of stiffener

When the outer edge of the stiffener is not continuously stiffened, the stiffener itself may buckle locally owing to the compressive force carried by the stiffener. In places of high compressive force, the edge of the stiffener is stiffened by using a T-section cut from rolled I-section or rolled angle section, with one side protruding outwards to serve as a stiffening member.

In our case, we are simply using two stiffener plates on either side of the web. The outstand of the stiffener plate should be limited in order to avoid local buckling of the stiffener. Referring to Table 5.2 (sheet 2) of Eurocode 3, Part 1-1, "Maximum width-to-thickness ratios for compression parts" (see Appendix B), we have the following.

For class 1 classification, the outstand c of the stiffener should not exceed $9t\varepsilon$, where

c = outstand of stiffener = width of stiffener-weld size,

t = thickness of stiffener = 30 mm (assumed),

$\varepsilon = (235/f_y)^{0.5} = (235/275)^{0.5} = 0.92,$

f_y = 275 N/mm^2 (for S 275 grade steel).

So, outstand of stiffener $c = 9 \times 30 \times 0.92 = 249$ mm.

An allowance should be made for the provision of a cope hole for welding. Assume that the size of the cope hole = 30 mm. Therefore, the maximum width of the stiffener plate is

b_s = 249 + 10 (assumed weld size) + 30 (cope hole) = 289, say 300 mm.

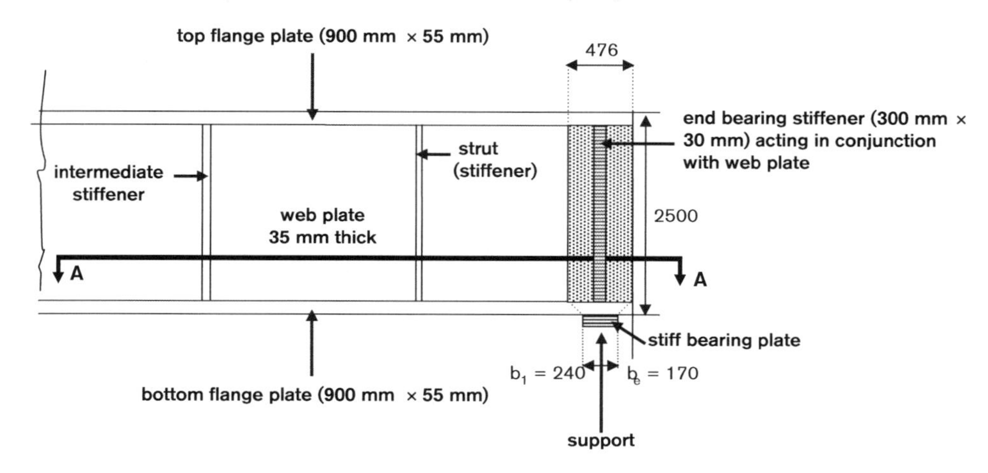

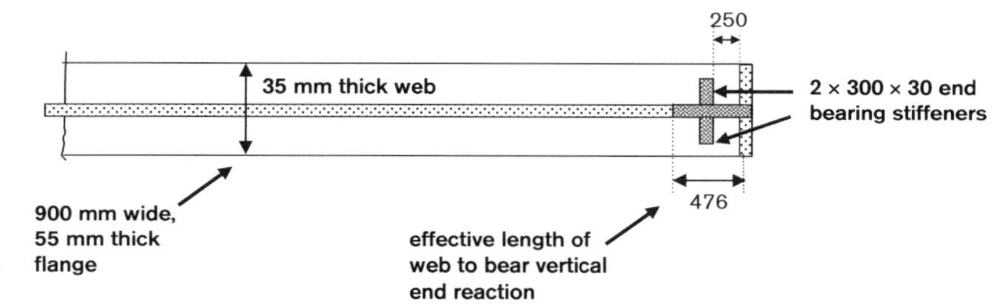

Fig. 3.6. End bearing stiffeners (shaded) acting in conjunction with web of a gantry plate girder to bear end reaction

Bearing capacity of stiffener

Therefore the bearing capacity of the end stiffener $P_s = A_{s,\,net}f_y$ where $A_{s,net}$ = net cross-sectional area of the stiffeners, allowing for cope holes for welding.

$$A_{s,net}\ \text{(effective)} = 2 \times (300 \times 30 - 30 \times 30) = 16\,200\ \text{mm}^2$$

$$f_y = 275\ \text{N/mm}^2$$

(for thickness of stiffener 30 mm). Therefore

$$\text{bearing capacity of stiffener} = P_s = 16\,200 \times 275/10^3 = 4455\ \text{kN}.$$

The net compressive force to be taken by the stiffener is

$F_{x\,net} = F_x - P_{bw} = 6281 - 3978 = 2303\ \text{kN} < P_s.$ <u>OK</u>

Total bearing capacity of (stiffener + web) = [4455 + 3978 (calculated previously)] kN = 8433 kN > 6281 kN (N_{Ed}). <u>Satisfactory</u>

3.3.9.3 Buckling resistance of load-bearing end stiffeners

The external load or reaction N_{Ed} (V_{vu}) on a load-carrying stiffener may cause the stiffener to buckle along with the web plate welded to it, and should not exceed the buckling resistance $N_{b,Rd}$ of the stiffener in conjunction with the web plate. This is given by the following equation:

$$N_{b,Rd} = (\chi A f_y)/\gamma_{M1} \quad (6.49)$$

where χ = reduction factor. To arrive at this reduction factor, we have to calculate the non-dimensional slenderness $\bar{\lambda}$. Referring to Clause 6.3.1.3, this is given by the equation

$$\bar{\lambda} = L_{cr}/(i\lambda_1) \quad (6.50)$$

where L_{cr} = buckling length in the buckling plane of stiffener

$= 0.7h_1 = 0.7 \times 2390 = 1693$ (h_1 = depth of web).

The factor of 0.7 is taken assuming that the stiffener ends are restrained.

A = area of cross-section of stiffener + area of web consisting of a certain length of web on either of stiffener × web thickness).

Grinter (1961) suggested that one should allow a length of web

l_1 = length of stiff bearing + distance from edge of stiff bearing to end of girder + 1/4 depth of girder,

assuming a 45° dispersion of load, giving

$$l_1 = b_1 + b_e + (1/4)h = 240 + 120 + 2500/4 = 985 \text{ mm}.$$

The BS code specifies 15 times the thickness of the web on either side of the stiffener, i.e.

$$l_1 = 15t_w + b_1/2 + b_e = 15 \times 35 + 240/2 + 170 = 815 \text{ mm}.$$

The Eurocode does not mention this. So, we assume that the effective length of web that acts with the stiffener is l_1 = 815 mm. Therefore

$A = 16\,200 + 815 \times 35 = 44725 \text{ mm}^2$,

I_{yy} = moment of inertia of combined section along web plane

$= 785 \times 35^3/12 + 30 \times 635^3/12 = 6.4 \times 10^8$,

r_{yy} = radius of gyration $= (I_y/A)^{0.5} = (6.4 \times 10^8/44\,725)^{0.5} = 1119.6$ mm,

$\lambda_1 = 93.9\varepsilon = 93.9 \times 0.96 = 90$.

Therefore

$$\bar{\lambda} = 1743/(145 \times 90) = 0.13 \text{ and } f_y = 275 \text{ N/mm}^2.$$

With the depth of the stiffener h_1 = 635 and the width of the web = 815 mm, welded section, referring to Table 6.2 of Eurocode 3, Part 1-1, "Selection of buckling curve for a cross-section", total depth of stiffener/thickness of stiffener = 635/30 = 21 < 30.

So, referring to Fig. 6.4 of Eurocode 3, Part 1-1, we follow the curve "c". With $\bar{\lambda} = 0.13$, $\chi = 1.0$. Therefore

$N_{b,Rd}$ = buckling resistance of end stiffener

$= 1.0 \times 44\,725 \times 255/10^3 = 11\,405$ kN $\gg N_{ed}$ (6281 kN). OK

3.4 Intermediate transverse stiffeners

3.4.1 Principles of the behaviour of intermediate stiffeners

As can be explained by the theory of principal stresses, the combination of the direct stress due to shear with the stress due to bending results in a compression in the web acting in the direction indicated by the dotted curved line in Fig. 3.7 (Spofford, 1938).

The curved lines of the principal stress in the web cut the neutral axis at 45°. Therefore, on the neutral axis the tension lines (as shown earlier in Fig. 3.4) cut the compression lines (shown in Fig. 3.6) at 45°. The maximum length of any diagonal strip between the stiffeners or across the web depth is either $s\sqrt{2}$ or $d\sqrt{2}$, whichever is the lesser. However, these strips do not behave simply as compression members. The strips tend to buckle outwards from the plane of the paper, but are restrained to a considerable extent by the diagonal strips at right angles to the strip under consideration. The usual practice to prevent buckling is to introduce pairs of plates or some other rolled sections as intermediate stiffeners that act as struts to carry the total vertical shear V in each section of the girder. A certain length of web also carries shear and behaves as a strut member. Assuming that the tension plane of the web is at an angle of 45°, the intermediate stiffeners should not be placed at more than one and half times the girder depth.

In our case, we assume that the stiffener is placed at a distance of 2400 mm from the end stiffener.

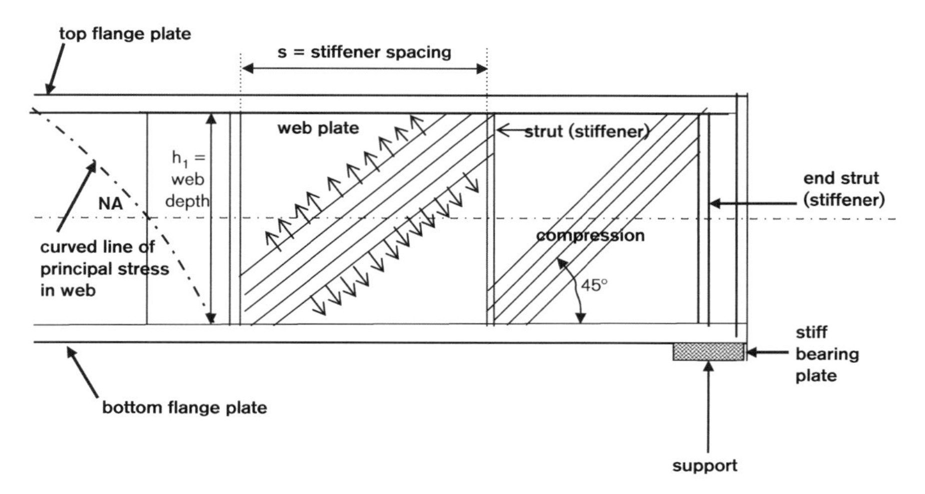

Fig. 3.7. Diagonal compression lines in the web plate of a plate girder

3.4.2 Design considerations

Two types of intermediate transverse web stiffeners are used:

(1) Stiffeners not subjected to a local concentrated load in the section considered.

(2) Stiffeners subjected to a local concentrated load in addition to a uniformly distributed load.

In our case, owing to the wheel loads, we shall design for a local point load in the section, with a front crane wheel load on that section.

To calculate the ultimate vertical shear force at 2.4 m from the right support, we do the following.

3.4.2.1 Ultimate shear due to wheel loads

We draw an influence line diagram at 2.4 m from the support for the shear force due to wheel loads with one wheel load at the section (see Fig. 3.8). From the influence line diagram,

$$\text{Sum of ordinates } \sum s = s_1 + s_2 + s_3 + s_4 + s_5 + s_5 + s_6 + s_7 + s_8$$
$$= (0.9 + 0.85 + 0.78 + 0.73 + 0.46 + 0.41 + 0.35 + 0.35) = 4.78.$$

The dynamic load from a wheel, for each of the eight wheels, is $c_k = 728$ kN, including a 40% dynamic factor, and the partial factor $\gamma_Q = 1.5$ for a moving wheel load. Therefore the maximum ultimate vertical shear force due to the wheels is

$$V_1 = \sum s c_k \gamma_Q = 4.78 \times 728 \times 1.5 = 5220 \text{ kN}.$$

3.4.2.2 Ultimate shear due to self-weight

Self-weight of girder/m = 11.92 kN/m; reaction $R_l = R_r = 11.92 \times 23.4/2 = 139.46$ kN.

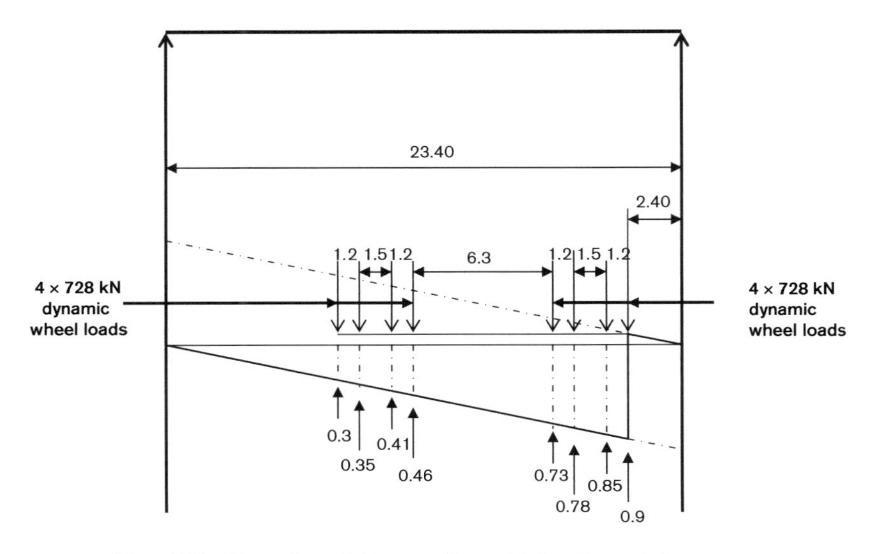

Fig. 3.8. Shear force influence line at 2.4 m from right support

With a partial factor $\gamma_G = 1.35$ for the self-weight (permanent load),

$$\text{ultimate shear at 2.4 m from the support} = V_2 = \gamma_G(R_r - 11.92 \times 2.4)$$
$$= 1.35 \times (139.46 - 11.92 \times 2.4) = 150 \text{ kN}.$$

3.4.2.3 Total ultimate shear at 2.4 m from support

$$\text{Total ultimate shear} = V_{Ed} = V_1 + V_2 = 5220 + 150 = 5370 \text{ kN}.$$

3.4.3 Design of intermediate stiffeners

The intermediate transverse stiffeners may be provided on either one side or both sides of the web. In our case, the stiffeners will be provided on both sides of the web.

3.4.3.1 Spacing of intermediate stiffeners

As already explained in Section 3.4.1, the spacing of the intermediate stiffeners should be limited to either $s\sqrt{2}$ or $d\sqrt{2}$, whichever is the lesser, in order for the stiffeners to act effectively as strut members. Thus, referring to Table 5.2 (sheet 1) of Eurocode 3, Part 1-1, the spacing of stiffeners should not exceed 1.5 times the depth of the web, provided that the ratio of the depth of the web to the thickness of the web does not exceed 72ε, where $\varepsilon = \sqrt{(235/f_y)}$. Thus, $s \leq 1.5d_1$ provided $d_1/t_w \leq 72\varepsilon$,

$$\text{where } h_1 = \text{height of stiffener} = h - 2t_f = 2500 - 2 \times 55 = 2390 \text{ mm},$$
$$72\varepsilon = 72 \times (235/f_y)^{0.5} = 72 \times (235/255)^{0.5} = 69.$$

In our case, $h_1/t_w = 2390/35 = 68.3 < 72\varepsilon\ (69)$ satisfies the condition.

$$\text{Maximum spacing that can be allowed} = 1.5 \times 2390 = 3585 \text{ mm}.$$

However, we have assumed a spacing $s = 2400$ mm. Therefore we must provide a spacing of intermediate stiffeners $s = 2400$ mm, which is greater than 2390 mm but less than 3585 mm. So, OK.

3.4.3.2 Outstand of intermediate stiffeners

When the outer edge of a stiffener is not continuously stiffened, the stiffener itself may buckle locally owing to the compressive force carried by the stiffener. To avoid local buckling of the edge, a T-section cut out from an I-section or angle section with one side protruding outwards to act as an edge stiffener is normally used when a stiffener is subjected to a high concentrated vertical load.

In our case, we shall use two stiffener plates on either side of the web plate. The outstand of the stiffener should be limited in order to prevent local buckling of the unstiffened plate. Referring to Table 5.2 ("Maximum width-to-thickness ratios for compression parts") of Eurocode 3, Part 1-1, for class 1 section classification, the outstand of the stiffener c should not exceed $t_s \times 9\varepsilon$, where c = outstand of stiffener, t_s = thickness of stiffener and $\varepsilon = \sqrt{(235/f_y)}$. Assuming a thickness of the stiffener plate $t_s = 25$ mm and $f_y = 275$ N/mm^2,

$$c = 25 \times 9 \times \sqrt{(235/275)} = 208 \text{ mm}.$$

Allowance should be made for the provision of a cope hole for welding between the web and the flange plate. Assuming that the size of the cope hole is 30 mm and size of the weld between web and stiffener is 10 mm,

maximum width of intermediate stiffener plate $b_s = 208 + 30 + 10 = 248$, say 250 mm (see Fig. 3.9).

3.4.3.3 Minimum stiffness of stiffener subjected to external applied loadings

The minimum stiffness of a stiffener *not* subjected to external loading is calculated as follows.

For $s/h_1 < \sqrt{2}$, $I_y = 1.5(h_1/s)^2 h_1 (t_{s\,min})^3$

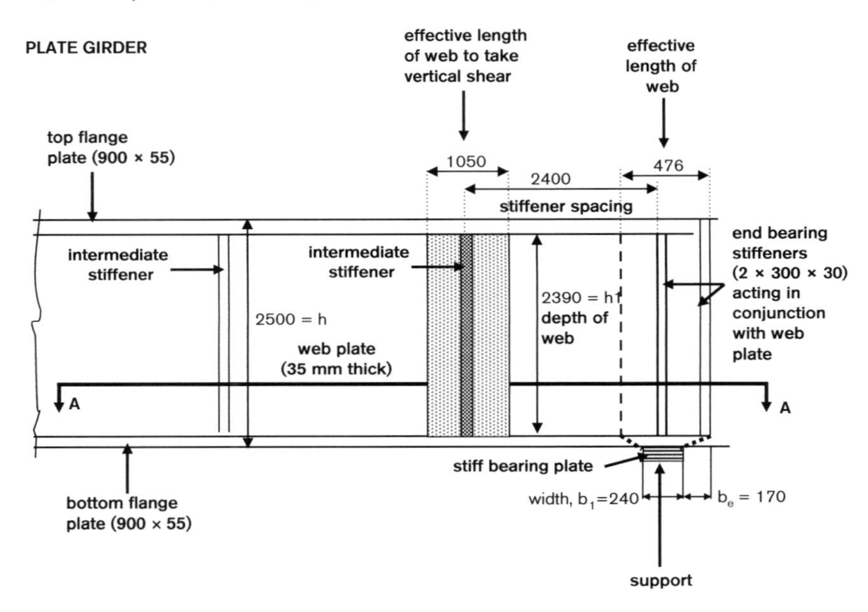

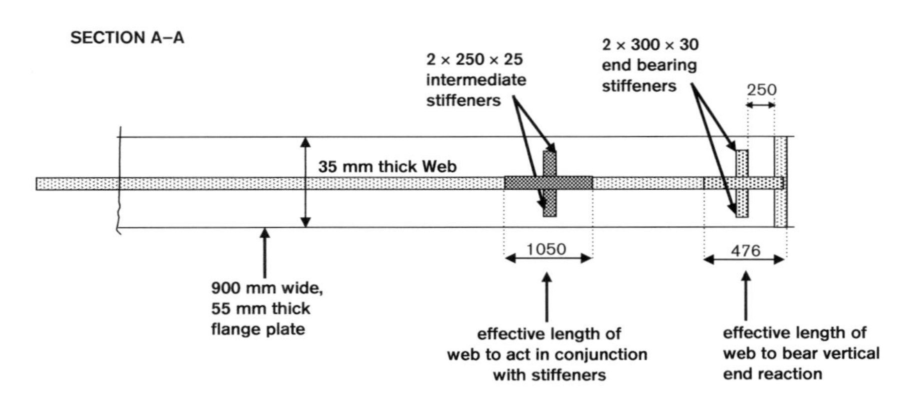

Fig. 3.9. Intermediate stiffeners acting in conjunction with the web (shaded) of a gantry plate girder to bear wheel load reaction

where I_y = moment of inertia along centre line of web and h_1 = height of stiffener = $h - 2t_f = 2390$.

$$\text{For } a/h_1 \geq \sqrt{2},\ I_y = 0.75d(t_{s\,min})^3$$

In our case,

$$s/h_1 = 2400/2390 = 1.04 < \sqrt{2}$$

Therefore

$$I_y = \text{minimum stiffness required} = 1.5 \times (2390/2400)^2 \times 2390 \times 25^3 = 5.6 \times 10^7\ \text{mm}^4.$$

Owing to the externally applied wheel load on the section, an additional stiffness is required.

- $I_{ext} = 0$ when the applied load is acting in line with the web, and
- $I_{ext} = V_{Ed}e_{yy}h^2/(Et_w)$

where E = elasticity modulus, e_{yy} = eccentricity of external transverse load from the centre line of the web and V_{Ed} = external transverse force (load).

In our case, there is no eccentricity developed. Therefore the minimum stiffness is required.

Normally, the actual stiffness is calculated with a portion of the web acting in conjunction with the stiffener plate. In practice, 15 times the thickness of the web on either side of the stiffener is taken into account for stiffener stiffness. This means that a length of web equal to

$$l_w = 2 \times 15t_w = 30 \times 35 = 1050\ \text{mm}$$

is acting along with the stiffener plates to provide the stiffness of the stiffener. Therefore

$$\text{actual stiffness of stiffener } Iyy\ (\text{actual}) = t_w(l_w - t_s)^3/12 + t_s \times (2b_s + t_w)^3/12$$
$$= 35 \times (1050 - 25)^3/12 + 25 \times (2 \times 250 + 35)^3/12 = 3.45 \times 10^9\ \text{mm}^4.\ \text{So, OK.}$$

3.4.3.4 Buckling resistance of stiffener

As already stated, the intermediate stiffeners are subjected to additional local crane wheel loads in addition to the uniformly distributed load of the girder.

Maximum ultimate shear in section 2.4 m from right support $V_{Ed} = 5370$ kN (calculated previously).

$$\text{Buckling resistance } N_{b,Rd} = (\chi A f_y)/\gamma_{M1} \quad (6.47)$$

where χ is the reduction factor. To arrive at this reduction factor, we refer to Clause 6.3.1.3, which gives the equation

$$\bar{\lambda} = L_{cr}/(i\lambda_1) \quad (6.50)$$

where L_{cr} = effective buckling height of stiffener in the buckling plane of the stiffener

$$= 0.7h_1 = 0.7 \times 2390 = 1673\ \text{mm}$$

(the ends are assumed restrained),

A = area of stiffeners + area of web acting with stiffener plates

$= 28(b_s - 30)ts + l_w t_w = 2 \times (250 - 30) \times 25 + 2 \times 1050 \times 35 = 52150 \text{ mm}^2,$

I_{yy} = moment of inertia = $3.45 \times 10^9 \text{ mm}^4$ (calculated before),

i_{yy} = radius of gyration = $(3.45 \times 10^9/52150)^{0.5} = 257$ mm,

$\lambda_1 = 93.9\varepsilon = 93.9 \times 0.96 = 90,$

$\bar{\lambda} = L_{cr}/(i_{yy}\lambda_1) = 1673/(257 \times 90) = 0.07.$

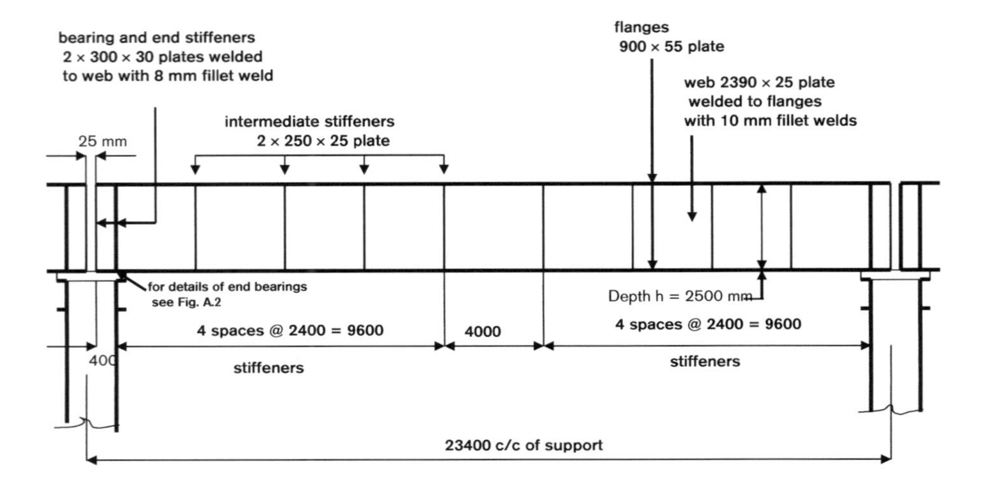

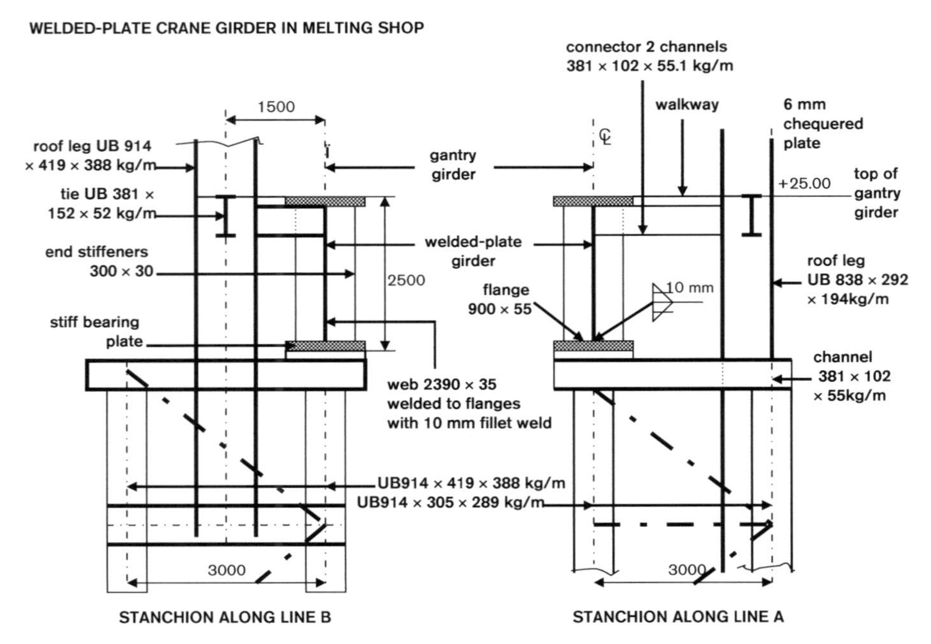

Fig. 3.10. Details of crane girder

Referring to Table 6.2 ("Selection of buckling curve for a cross-section") of Eurocode 3, Part 1-1, with an effective depth of stiffener $h_s = 2 \times 250 + 35 = 535$ and thickness of stiffener $t_s = 25$,

$$h_s/t_s = 535/25 = 21 \text{ and } f_y = 275 \text{ N/mm}^2.$$

Referring to Table 6.2 of Eurocode 3, Part 1-1 again to select the buckling curve "c" in Fig. 6.4 of that Eurocode, the reduction factor $\chi = 1.0$. Then,

buckling resistance $N_{b,Rd} = \chi A f_y/\gamma_{M1} = 1 \times 52\,150 \times 275/10^3 = 14\,341$ kN $\gg V_{Ed}$ (5370 kN). Satisfactory

3.5 Design of end bearings of gantry girder

For the design and details of the bearings, see Appendix A. For details of the crane girder, see Fig. 3.10.

References

British Standards Institution, 1983. BS 2573-1: 1983, Rules for the design of cranes. Specification for classification, stress calculations and design criteria for structures.

British Standards Institution, 1984. BS 466: 1984, Specification for power driven overhead travelling cranes, semi-goliath and goliath cranes for general use.

Eurocode, 2002. BS EN 1990: 2002(E), Basis of structural design.

Eurocode, 2005. BS EN 1993-1-1: 2005, Eurocode 3. Design of steel structures. General rules and rules for buildings.

Eurocode, 2006. BS EN 1991-3: 2006, Actions on structures. Actions induced by cranes and machines.

Grinter, L.E., 1961. *Design of Modern Steel Structures*, Macmillan, New York.

Salmon, E.H., 1948. *Materials and Structures*, Longmans, Green & Co., London.

Spofford, C.M., 1939. *The Theory of Structures*, McGraw-Hill, New York.

Stewart, D.S., 1953. *Practical Design of Simple Steel Structures*, Constable, London.

Timoshenko, S.P. and Gere, J.M., 1961. *Theory of Elastic Stability*, McGraw-Hill Kogakusha, Tokyo.

CHAPTER 4

Design of Welded and Bolted Connections

The design of welded and bolted connections described here is based on Eurocode 3, Part 1-8 (Eurocode, 2005a). All of the references to clauses and equations (referred to as "expressions" in the Eurocode) in this chapter relate to the above Eurocode unless otherwise stated.

4.1 General

Joints should be designed based on realistic assumptions about the distribution of internal forces. These assumptions should be such that in all cases the load transfer will take place directly through the joint relative to the stiffnesses of the various components of the joint, and the internal forces and external applied forces should be in equilibrium. Ease of fabrication and erection should be taken into account in the design of connections.

4.1.1 Joints in simple design

In simple design, the joints between members should have adequate capacity for transmitting the calculated forces and should not develop any significant moments that may adversely affect the structural strength of members.

4.1.2 Joints in continuous design

In continuous design, joints between members should have adequate capacity for transmitting the forces and moments calculated in a global analysis. In the case of an elastic analysis, the rigidity of the joints should be such that the stiffness of the frame is not less than that assumed in the analysis to such an extent that this would reduce the load-carrying capacity.

4.1.3 Method of connection

The connections can be effectively achieved using either bolts or welds.

4.2 Welded connections

Welded connections are mainly of two types, namely fillet welds and butt welds.

In *fillet weld* connections, the connecting members are joined to the parent member either directly or using splice plates by means of fillet welding. For example, fillet welds are used

- when the columns of stanchions are connected to the base or cap by means of fillet welding;

- when lengths of columns or beams are spliced together by splice plates welded onto flanges.

In *butt welds*, two connecting members are welded directly end to end by means of full-penetration butt welds to achieve the same strength as that of the parent members. For example, butt welds are used when columns or beams are butt welded end to end at splice points during extension, rather than joined by splice plates.

4.2.1 Design of fillet welds

4.2.1.1 Size of fillet welds

The size of a fillet weld, denoted by the letter s, is the length of the leg for a plain fillet weld on the fusion surface of the parent metal.

4.2.1.2 Throat of fillet welds

The throat of a fillet weld, denoted by a, is the perpendicular distance from the root of the weld to a straight line joining the fusion faces that lie within the cross-section of the weld. So,

$$\text{effective throat thickness} = a = 0.7s$$

(assuming that the straight line joining the fusion surfaces is at 90°), and in the design, the strength of a fillet weld is always calculated based on the throat thickness.

4.2.1.3 Welding consumables

In accordance with Eurocode 3, Part 1-8, the design strength of a weld depends on the steel grade of the parent metal and the class of electrodes (welding consumables) used. All welding consumables should conform to the relevant standards specified in Clause 1.2.5 of the Eurocode, "Reference standards, Group 5: Welding consumable and welding" (see also BS EN ISO 14555: 1998, BS EN 12345: 1999 and BS EN 13918: 2003 (British Standards Institution, 1998, 1999, 2003)). The specified yield strength, ultimate tensile strength, elongation at failure and minimum Charpy V-notch energy value of the filler metal should be equivalent to or better than that specified for the parent material. Normally it is safe to use electrodes that are overmatched with regard to the steel grades used.

Table 4.1 gives values of the design strength $p_w = f_{vw.d}$ of a weld for different grades of steel and different classes of electrodes used in the welding operation.

4.2.1.4 Design resistance of fillet welds

Referring to Clause 4.5.3.3, "Simplified method for resistance of fillet weld", we have the following.

Table 4.1. Design strength of fillet welds $f_{vw.d}$

Steel grade	Electrode classification		
	35	42	50
S 275	220 N/mm^2	220 N/mm^2 [a]	220 N/mm^2 [a]
S 355	220 N/mm^2 [b]	250 N/mm^2	250 N/mm^2 [a]
S 460	220 N/mm^2 [b]	250 N/mm^2 [b]	250 N/mm^2

[a] Overmatching electrodes.
[b] Undermatching electrodes.

The design resistance of a fillet weld may be assumed to be adequate if, at every point along its length, the resultant of all the forces per unit length transmitted by the weld satisfies the following criterion (see Eurocode 3, Part 1-8):

$$F_{w,Ed} \leq F_{w,Rd} \quad (4.2)$$

where $F_{w,Ed}$ is the design value of the weld force per unit length and $F_{w,Rd}$ is the design weld resistance per unit length.

Independent of the orientation of the weld throat plane relative to the applied force, the design resistance per unit length $F_{w,Rd}$ is given by the following equation:

$$F_{w,Rd} = f_{vw.d}a \quad (4.3)$$

where $f_{vw.d}$ is the design shear strength of the weld and a is the throat thickness of the weld.

The design shear strength $f_{bw.d}$ of the weld is given by the following equation:

$$f_{vw.d} = f_u/\sqrt{3}/(\beta_w \gamma_{M2}) \quad (4.4)$$

where f_u is the nominal ultimate tensile strength of the weaker part joined, taken from Table 4.2; γ_{M2} is the resistance of joints, equal to 1.25 (see Eurocode 3, Part 1-1); and β_w is an appropriate correlation factor, taken from Table 4.3.

Table 4.2. Nominal values of yield strength f_y and ultimate tensile strength f_u for hot-rolled structural steel (based on Table 3.1 of Eurocode 3, Part 1-1 (Eurocode, 2005b), using standard EN 10025-2)

	Nominal thickness t of element (mm)			
	$t \leq 40$ mm		40 mm $< t \leq 80$ mm	
Steel grade	f_y (N/mm^2)	f_u (N/mm^2)	f_y (N/mm^2)	f_u (N/mm^2)
S 235	235	360	215	360
S 275	275	430	255	410
S 355	355	510	335	470
S450	440	550	410	550

Table 4.3. Correlation factor β_W for fillet welds (based on Table 4.1 of Eurocode 3, Part 1-8)

Standard and steel grade	Standard and steel grade	Standard and steel grade	Correlation factor β_w
EN 10025	EN 10210	EN 10219	–
S 235, S 235W	S 235H	S 235H	0.8
S 275, S 275N/NL, S 275M/ML	S 275H, S 275NH/ NLH	S 275H, S 275NH/NLH, S 275MH/MLH	0.85
S 355, S 355N/NL, S 355M/ML, S 355W	S 355H, S 355NH/ NLH	S 355H, S 355NH/NLH, S 355MH/MLH	0.9
S 420 N/NL, S 420M/ML	–	S 420MH/MLH	1.0
S 460 N/NL, S 460M/ML, S 460Q/QL/QL1	S 460 NH/NLH	S 460 NH/NLH, S 460MH/MLH	1.0

4.2.1.5 Design of weld between flange and web of plate girder in melting bay (see Fig. 3.9)

Design assumptions

- The connection is subjected to moments, shears and thrusts due to crane loads. The weld connection should be designed to resist the moments and forces.
- Owing to moments, the weld will undergo tensile or compressive bending stresses in the throat plane. So, if $M_u = M_{Ed}$ is the ultimate bending moment acting in the welded connection and Wy is the elastic modulus of the weld group in the section considered, then the bending stress is $f_{wb} = M_{Ed}/W_y$.
- Owing to shear in the weld, the weld will develop a shear stress in the throat plane. So, if $F_{w,Ed}$ is the ultimate shear in the weld throat plane per unit length and A_w is the shear area of the weld throat plane per unit length, then the shear stress is $f_{wv} = F_{w,Ed}/A_w$.
- Owing to the direct tensile force on the weld, the weld will develop a tensile stress in the weld throat plane. So, if F_t is the tensile force on the weld and A_w is the area of the weld throat plane, then the tensile stress is $f_{wt} = F_t/A_w$.

As the direction of the shear stress is perpendicular to the direction of the bending stress and tensile stress, the resultant stress on the web will be the vector sum of the above. Thus,

resultant stress on the web $f_{wr} = [(f_{wb} + f_{wt})^2 + f_{wv}{}^2]^{0.5}$.

Design moments and forces

Consider the maximum ultimate design moment at 10.1 m from the left support, $M_{Ed} = M_{vu} = 29\ 515$ kN m (previously calculated, see Chapter 3), the maximum ultimate design shear at the support, $V_{Ed} = 5882$ kN, and the maximum ultimate design shear at 10.1 m from the left support,

$$V_{Ed,10.1} = \gamma_{crk} V = 1.5 \times 1645 = 2468 \text{ kN}$$

where γ_{Crk} is the partial safety factor for moving crane loads, equal to 1.5, and V is the characteristic shear at 10.1 m from the left support, calculated from the shear influence line diagram.

Design of weld size

Horizontal shear stress f_h/linear length = vertical shear stress f_v/linear depth.

So, we shall calculate the maximum horizontal shear stress due to the maximum vertical ultimate shear. By an approximate formula,

vertical shear/linear height of girder = V_{Ed}/h_w

Now,

horizontal shear/linear length = vertical shear/linear height of girder,

where

h_w = depth between top and bottom of welds of
girder = $(h - 2t_f) = 2500 - 2 \times 55 = 2390$ mm.

Therefore the design horizontal shear/linear length of the weld is

$$F_{w,Ed} = V_{Ed}/2390 = 5882/2390 = 2.46 \text{ kN}.$$

The design shear strength is

$$f_{vw.d} = f_u/\sqrt{3}/(\beta_w \gamma_{M2}) \tag{4.4}$$

From Table 4.3, $f_u = 430$ N/mm^2 for steel grade S 275.

Referring to Clause 6.1 of Eurocode 3, Part 1-1, the partial factor for resistance for joints γ_{M2} is 1.25. Therefore

$$f_{vw.d} = 430/(3)^{0.5}/(0.85 \times 1.25) = 234 \text{ N/mm}^2.$$

Assume that the size of the weld $s = 10$ mm and the throat thickness $a = 0.7 \times 10 = 7$ mm. Therefore

$$\text{weld resistance/unit length} = F_{w,Rd} = f_{vw.d} \times a \times 2 = 234 \times 7 \times 2/10^3$$
$$= 3.28 \text{ kN} > F_{w,Ed} \ (2.46).$$

By an exact formula,

$$\text{horizontal shear } F_{w,Ed} = V_{Ed}G/I,$$

where G is the first moment (gross area of flange × distance) about the neutral axis of all of the area above the horizontal layer considered (i.e. above the web surface), given by

$$G = bt_f(h_w/2 + t_f/2)$$
$$= 900 \times 55 \times (2390/2 + 27.5) = 60\,513\,750 \text{ mm}^3.$$
$$I = \text{moment of inertia of the total gross section in mm}^4$$
$$= t_w h_w^3/12 + 2bt_f(h_w/2 + t_f/2)^2$$
$$= 35 \times 2390^3/12 + 2 \times 900 \times 55 \times (2390/2 + 55/2)^2 = 1.88 \times 10^{11} \text{ mm}^4.$$

Therefore

horizontal shear = vertical shear;

$$F_{w,Ed} = V_{Ed}G/I = 5882 \times 10^3 \times 60\,513\,750/(1.88 \times 10^{11}) = 1893 \text{ N/linear mm}.$$

Therefore

$$\text{horizontal shear/linear mm} = F_{w,Ed} = 1893 \text{ N/linear mm}.$$

Assume that the size of the fillet weld s is 10 mm and the throat thickness a is $0.7 \times 10 = 7$ mm as before. With a design strength of the fillet weld $f_{vw.d} = 234$ N/mm^2 (calculated above), the design shear resistance of the weld on both sides of web per linear millimetre is

$$F_{w,Rd} = 234 \times 7 \times 2 = 3276 \text{ N} > F_{w,Ed} \ (1893 \text{ N}).$$

In addition, we have to consider the local vertical shear on the weld due to the wheel on the flange.

Dynamic wheel load = 728 kN.

Referring to Table A1(B) of BS EN ISO 14555: 1998 (British Standards Institution, 1998), with a partial safety factor $\gamma_{Crk} = 1.5$ for variable crane loads,

ultimate wheel load = 728 × 1.5 = 1092 kN,

spacing of wheels = 1.2 m.

Therefore

equivalent uniform wheel load/m = 1092/1.2 = 910 kN/m;

equivalent uniform vertical shear in N/mm = F_v = 910 × 1000/1000 = 910 N/mm.

Therefore the resultant shear on the weld per linear length is

$F_{Rw,Ed} = \sqrt{(F_{w,Ed}^2 + F_v^2)} = \sqrt{(1893^2 + 910^2)} = 2100$ N/linear mm.

Shear capacity of 10 mm weld (as calculated above) = $F_{w,Rd} = 3276$ N > F_r (2100 N)/mm.

Therefore we adopt 10 mm welds (class 35 electrode) to connect the flange plates to the web plate (see Fig. 4.1).

4.2.1.6 Design of fillet weld between web and end stiffener of crane girder in melting bay (see Fig. 3.10)

Design assumptions

It is assumed that the end reaction from the crane girder is transmitted to the web by means of the end stiffener. There is one pair of end stiffeners. Each will carry half the end reaction.

Loads on stiffeners

End reaction = F_v = 5882 kN.

Depth of girder between flanges = d = 2500 − 2 × 55 = 2390 mm.

Vertical shear to be transmitted by the pair of stiffeners to the web per linear depth = 5882 × 10^3/2390 = 2461 N/linear mm depth.

Shear to be transmitted by each stiffener = $F_{w,Ed}$ = 2461/2
= 1232 N/linear mm depth.

Design of weld size

Assume a size of weld s = 8 mm and a throat thickness a = 0.7 × 8 = 5.6 mm. Assume that the design strength of the fillet weld is $f_{vw.d}$ = 234 N/mm^2 (calculated before). Therefore the design weld resistance of a 6 mm fillet weld and stiffener on both sides of the contact to the web per linear mm depth is

$F_{w,Rd}$ = 5.6 × 234 × 2 = 2621 N > (1232 N). Satisfactory

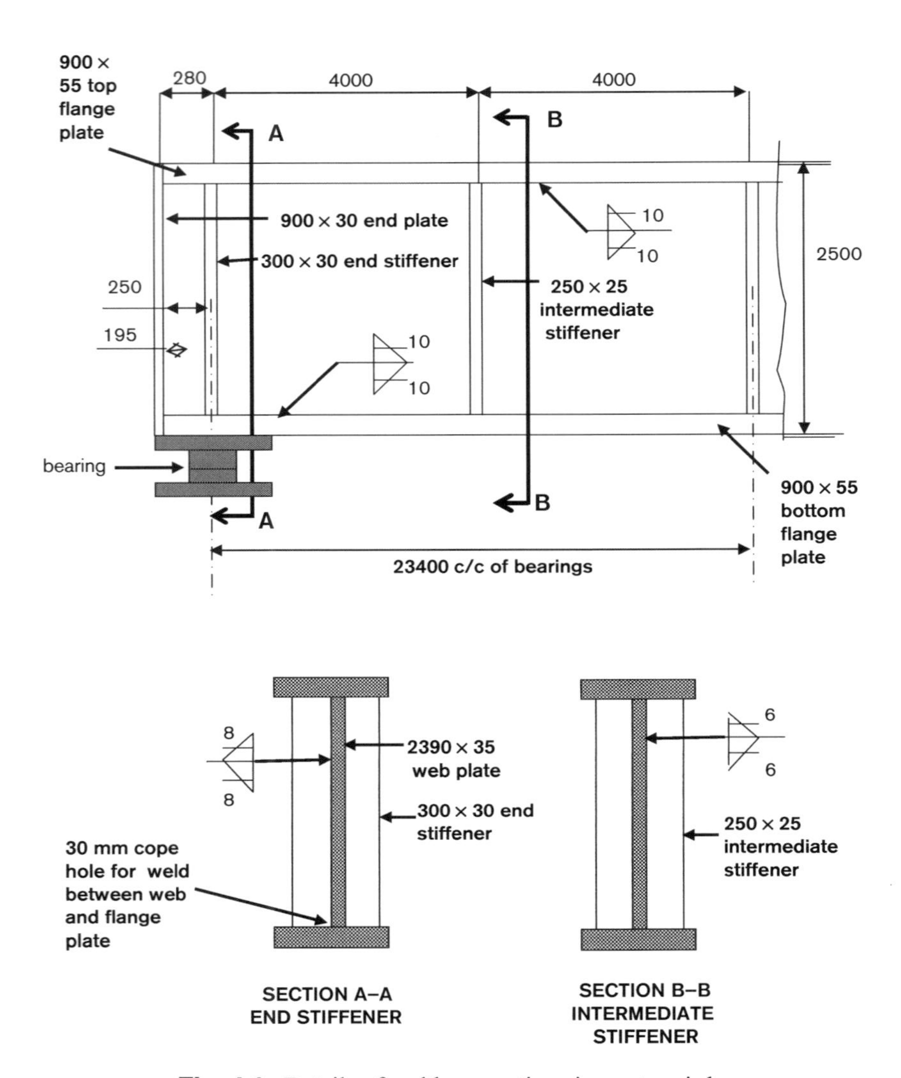

Fig. 4.1. Details of weld connections in gantry girder

Therefore we provide 8 mm fillet welds between the stiffener and the web for the full depth of the girder. The end of the web should be V-notched to receive the fillet weld between the web and the outer side of the end stiffener. For the intermediate stiffeners, we provide a 6 mm fillet weld.

4.2.1.7 Design of fillet welds between roof and crane legs and between gusset plates and base plate in the base of stanchion A (see Fig. 6.4)

Design assumptions

The crane leg is subjected to an ultimate compression $N_{c,Ed} = 9970$ kN and the roof leg is subjected to an ultimate tension $N_{t,Ed} = 3279$ kN (calculated in Chapter 6). We assume that both of these forces are transferred directly to the base through the weld connections between the gusset plates and flanges and between the web and base plate.

Design of size of welds

Assume that the size of weld $s = 15$ mm and the throat of the weld $a = 0.7 \times 15 = 10.5$ mm.

Design shear strength of weld $f_{vw.d} = 234$ N/mm^2 (calculated earlier).

Design weld resistance/linear mm of weld $F_{w,Rd} = 234 \times 10.4 = 2434$ N/mm.

Length of weld in contact between gusset plates and flanges and between web and base plate $l = 4 \times 650 + 2 \times 825 = 4250$ mm.

Total design resistance of weld $= N_{c,Rd} = 2434 \times 4250/10^3$
$= 10\,345$ kN $> N_{c,Ed}$ (9970 kN). Satisfactory.

Therefore we provide a 15 mm fillet weld between the stanchion legs and the gusset and base plates.

4.3 Design of bolted connections

4.3.1 Design assumptions

- The connections are designed to transfer moment, shear and thrust at the joint.
- The connections should be designed to the full strength (i.e. full capacity) of the members to be connected at the joint.
- In the case of a shallow member, it is the general practice to transfer the tensile and compressive forces in the flanges due to moments through double splice plates in the flanges, and to transfer the shear force in the web by double splice plates in the web.
- In the case of deep beams and plate girders, the connections are subjected to moment and shear.

The web plate splices can be calculated by the following two methods:

- Method 1: the flanges are assumed to take all the bending moment and the web is supposed to take only the shear. The web splice is designed for shear only.
- Method 2: the total moment in the section of the connection is shared by the web and the flanges. It is assumed that 1/8 of the web area will share the total BM. Thus,

BM shared by web = total BM × (1/8 of web area/total web area).

So, the web splice is designed for a BM as calculated above in addition to the full shear.

- The flange plate splices are designed for the full strength (effective) of the member to be connected.
- Normally, in moment connections, preloaded high-strength friction grip (HSFG) bolts are used to avoid slip in the joint, which could lead to an unacceptable deflection or change of moment, causing collapse of the member.

4.3.2 General requirements

All joints should have a design resistance such that the structure is capable of satisfying all of the basic design requirements given in Eurocode 3, Parts 1-1 and 1-8. The partial safety factors γ_M are given in Table 4.4 (based on Table 2.1, "Partial safety factors for joints", of Eurocode 3, Part 1-8).

Table 4.4. Partial safety factors for joints (based on Table 2.1 of Eurocode 3, Part 1-8)

Resistance of members and cross-sections	γ_{M0}, γ_{M1} and γ_{M2}: see Eurocode 3, Part 1-1
Resistance of bolts, resistance of plates in bearing	$\gamma_{M2} = 1.25$, recommended value
Slip resistance in ultimate limit state (Category C)	$\gamma_{M3} = 1.25$, recommended value
Slip resistance in serviceability limit state (Category B)	$\gamma_{M3,ser} = 1.1$, recommended value
Preload of high-strength bolts	$\gamma_{M7} = 1.1$, recommended value

4.3.3 Joints loaded in shear subject to impact, vibration and/or load reversal

Where a joint loaded in shear is subject to impact or significant vibration, one of the following jointing methods should be used:

- welding;
- bolts with locking devices;
- preloaded bolts.

Where slip is not acceptable in a joint because it is subject to reversal of shear load or for any other reason, preloaded bolts in a Category B or C connection (see Table 4.6), fit bolts (Clause 3.6.1 of Eurocode 3, Part 1-8) or welding should be used.

For wind and/or stability bracings, bolts in Category B or C connections are preferred.

4.3.4 Connections made with bolts

- All bolts, nuts and washers should comply with Clause 1.2.4, "Reference standards: Group 4".
- The rules in the standard are valid the for bolt classes given in Table 4.5.
- The yield strength f_{yb} and the ultimate strength f_{ub} for bolt classes 4.6, 4.8, 5.6, 5.8, 6.8, 8.8 and 10.9 are given in Table 4.5. These values should be adopted as characteristic values in design calculations.

4.3.5 Preloaded bolts (HSFG)

Only bolt assemblies of class 8.8 and 10.9 conforming to the requirements given in Clause 1.2.4, "Reference standards: Group 4", for high-strength structural bolting for preloading with controlled tightening in accordance with the requirements in Clause 1.2.7, "Reference Standards: Group 7", may be used as preloaded bolts. (See BS EN 1090-2 (British Standards Institution, 2008)).

4.3.6 Categories of bolted connections

There are two types of bolted connections, namely shear connections and tension connections.

Table 4.5. Nominal values of the yield strength f_{yb} and the ultimate tensile strength f_{ub} for bolts (based on Table 3.1 of Eurocode 3, Part 1-8)

Bolt class	4.6	4.8	5.6	5.8	6.8	8.8	10.9
f_{yb} (N/mm^2)	240	320	300	400	480	640	900
f_{ub} (N/mm^2)	400	400	500	500	600	800	1000

4.3.6.1 Shear connections

Bolted connections loaded in shear should be designed as one of the following:

- *Category A: bearing type.* Bolts from classes 4.6 to 10.9 should be used in this category. No preloading or special provisions for contact surfaces are needed. The design ultimate shear load should not exceed the design shear resistance calculated from the equations given in Clause 3.6.
- *Category B: slip-resistant in serviceability limit state.* Preloaded bolts in accordance with Clause 3.1.2(1) should be used in this category. Slip should not occur in the serviceability limit state. The design serviceability shear load should not exceed the design slip resistance, given in Clause 3.9. The design ultimate shear load should not exceed the design shear resistance, given in Clause 3.6, nor the design bearing resistance, also given in Clause 3.6.
- *Category C: slip-resistant in ultimate limit state.* Preloaded bolts in accordance with Clause 3.1.2(1) should be used in this category. Slip should not occur in the ultimate limit state. The design ultimate shear load should not exceed the design slip resistance, given in Clause 3.9, nor the design bearing resistance, given in Clause 3.6. In addition, for a connection, the design plastic resistance $N_{net,Rd}$ of the net cross-section at bolt holes (see Clause 6.2 of Eurocode 3, Part 1-1) should be checked in the ultimate limit state.

4.3.6.2 Tension connections

Bolted connections loaded in tension should be designed as one of the following:

- *Category D: non-preloaded.* Bolts from classes 4.6 to 10.9 should be used in this category. No preloading is required. This category should not be used where the connections are frequently subjected to variations of tensile loading. However, it may be used in connections designed to resist normal wind loads.
- *Category E: preloaded.* Preloaded class 8.8 and 10.9 bolts with controlled tightening in conformity with Clause 1.2.7, "Reference Standards: Group 7" should be used in this category.

The design checks for these connections are summarized in Table 4.6, where $F_{v,Ed}$ is the design shear force per bolt in the ultimate limit state, $F_{t,Ed}$ is the design tensile force per bolt in the ultimate limit state, $F_{v,Rd}$ is the design shear resistance per bolt, $F_{b,Rd}$ is the design bearing resistance per bolt, $F_{s,Rd,ser}$ is the design slip resistance per bolt in the serviceability limit state, $F_{s,Rd}$ is the design slip resistance per bolt in the ultimate limit state, $N_{net,Rd}$ is the design tension resistance of the net section at holes, $F_{t,Rd}$ is the design tension resistance per bolt and $B_{p,Rd}$ is the design punching shear resistance of the bolt and the nut.

4.3.7 Positioning of holes for bolts and rivets

The minimum and maximum spacings and the end and edge distances for bolts and rivets are given in Table 4.7, where e_1 is the distance from the centre of a fastener hole to the adjacent end of any part, measured in the direction of load transfer, e_2 is the distance from the centre of a fastener hole to the adjacent edge of any part, measured at right angles to the direction of load transfer, p_1 is the spacing between the centres of the fasteners in a line in the direction of load transfer, p_2 is the spacing measured perpendicular to the load transfer direction between adjacent lines of fasteners, d_0 is the hole diameter for a bolt, a rivet or a pin, and t is the thickness of the plate.

For the minimum and maximum spacings and the end and edge distances for structures subjected to fatigue, see BS EN 1993-1-9: 2005 (Eurocode, 2005c).

Table 4.6. Categories of bolted connections (based on Table 3.2 of BS EN 1993-1-8: 2005(E)) [a]

Category	Criterion	Remarks
Shear connections		
A, bearing type	$F_{v,Ed} \leq F_{v,Rd}$	No preloading required
	$F_{v,Ed} \leq F_{b,Rd}$	Bolt classes from 4.6 to 10.9 may be used
B, slip-resistant at serviceability	$F_{v,Ed} \leq F_{s,Rd,ser}$	Preloaded class 8.8 or 10.9 bolts should be used
	$F_{v,Ed} \leq F_{v,Rd}$	For slip resistance at serviceability, see Clause 3.9
	$F_{v,Ed} \leq F_{b,Rd}$	
C, slip-resistant at ultimate	$F_{v,Ed} \leq F_{s,Rd}$	Preloaded 8.8 or 10.9 bolts should be used
	$F_{v,Ed} \leq F_{b,Rd}$	For slip resistance at ultimate, see Clause 3.9
	$F_{v,Ed} \leq N_{net,Rd}$	For $N_{net,Rd}$, see Clause 3.4.1(1)c
Tension connections		
D, non-preloaded	$F_{t,Ed} \leq F_{t,Rd}$	No preloading required
	$F_{t,Ed} \leq B_{p,Rd}$	Bolt classes from 4.6 to 10.9 may be used
E, preloaded	$F_{t,Ed} \leq F_{t,Rd}$	Preloaded class 8.8 or 10.9 bolts should be used
	$F_{t,Ed} \leq B_{p,Rd}$	For $B_{p,Rd}$, see Table 3.4 of Eurocode 3, Part 1-8

[a] The design tensile force $F_{t,Ed}$ should include any force due to prying action; see Clause 3.11. Bolts subjected to both shear force and tensile force should also satisfy the criteria given in Table 3.4 of Eurocode 3, Part 1-8.

Table 4.7. Minimum and maximum spacing, end and edge distances (based on Table 3.3 of BS EN 1993-1-8: 2005(E))

		Maximum		
		Structures made from steels conforming to EN 10025, except steel conforming to EN 10025-5		Structures made from steels conforming to EN 10025-5
Distances and spacings	Minimum	Steel exposed to the weather or other corrosive influences	Steel not exposed to the weather or other corrosive influences	Steel used unprotected
End distance e_1	$1.2d_0$	$4t + 40$ mm	–	The larger of $8t$ or 125 mm
Edge distance e_2	$1.2d_0$	$4t + 40$ mm	–	The larger of $8t$ or 125 mm
Spacing p_1	$2.2d_0$	The smaller of $14t$ or 200 mm	The smaller of $14t$ or 200 mm	The smaller of $14t_{min}$ or 175 mm
Spacing p_2	$2.4d_0$	The smaller of $14t$ or 200 mm	The smaller of $14t$ or 200 mm	The smaller of $14t_{min}$ or 175 mm

4.3.8 Properties of slip-resistant connections using class 8.8 or 10.9 HSFG bolts and splice plates

- For HSFG bolts, the design shear strength of the bolt is achieved by preloading the bolt, with the development of an effective frictional resistance between the contact surfaces of the splices and the member to be connected.

- The HSFG bolts must conform to Clause 1.2.4, "Reference standards, Group 4: Bolts, nuts and washers".
- The strength grade should be bolt class 8.8 or 10.9 for high-strength bolts up to M24 diameter. For higher diameters of bolts, the ultimate strength of the bolt should be reduced.
- In connections using HSFG bolts, the thickness of the splice plates must not be less than half the diameter of the bolts or a minimum of 10 mm.

4.3.9 Design resistance of individual fasteners (HSFG bolts in preloaded condition)

The design resistance for individual fasteners subjected to shear and/or tension is given in Table 4.8, where A is the gross cross-sectional area of the bolt, A_s is the tensile area of the bolt, f_{ub} is the ultimate tensile strength for bolts, t is the thickness of the plate, d_0 is the hole diameter for a bolt, a rivet or a pin, d_m is the mean of the across-points and across-flats dimensions of the bolt head or nut, whichever is the smaller, and p_1 is the spacing between the centres of the fasteners in a line in the direction of load transfer.

For preloaded bolts, in accordance with Clause 3.1.2(1), the design preload $F_{p,Cd}$ to be used in design computations is given by the following equation:

Table 4.8. Design resistance for individual fasteners subjected to shear and/or tension (based on Table 3.4 of Eurocode 3, Part 1-8)

Failure mode	Bolts	Rivets
Shear resistance per shear plane	$F_{v,Rd} = \alpha v f_{ub} A/\gamma_{M2}$ where the shear plane passes through the threaded portion of the bolt (A is the tensile stress area of the bolt, equal to A_s). For classes 4.6, 5.6 and 8.8, $\alpha_v = 0.6$. For classes 4.8, 5.8, 6.8 and 10.9, $\alpha_v = 0.5$. Where the shear plane passes through the unthreaded portion of the bolt (A is the gross cross-sectional area of the bolt), $\alpha_v = 0.60$.	$F_{v,Rd} = 0.6 f_{ur} A_0/\gamma_{M2}$
Bearing resistance	$F_{b,Rd} = k_1 a_b f_u dt/\gamma_{M2}$ where a_b is the smallest of α_d, f_{ub}/f_u or 1.0 in the direction of load transfer. For end bolts, $\alpha_d = e_1/3d_0$. For inner bolts, $\alpha_d = p_1/3d_0 - 1/4$ perpendicular to the direction of load transfer. For edge bolts, k_1 is the smaller of $2.8e_2/d_0 - 1.7$ or 2.5. For inner bolts, k_1 is the smaller of $1.4p_2/d_0 - 1.7$ or 2.5.	–
Tension resistance	$F_{t,Rd} = k_2 f_{ub} A_s/\gamma_{M2}$ where $k_2 = 0.63$ for countersunk bolt; otherwise $k_2 = 0.9$.	$F_{t,Rd} = 0.6 f_{ur} A_0/\gamma_{M2}$
Punching shear resistance	$B_{p,Rd} = 0.6\pi d_m t_p f_u/\gamma_{M2}$	No check needed
Combined shear and tension	$F_{v,Ed}/F_{v,Rd} + F_{t,Ed}/1.4F_{t,Rd} \leq 1$	–

$$F_{p,Cd} = \text{design preload} = 0.7 f_{ub} A_s / \gamma_{M7} \quad (3.1)$$

where f_{ub} is the ultimate tensile strength of the bolt, A_s is the tensile stress area of the bolt and γ_{M7} is the partial safety factor for bolts, equal to 1.1.

The design resistance for tension and shear through the threaded portion of a bolt given in Table 4.8 should be used only for bolts manufactured in accordance with the reference standard group described in Clause 1.2.4.

The design shear resistance $F_{v,Rd}$ given in Table 4.8 should be used only where the bolts are used in holes with nominal clearances not exceeding those for normal holes as specified in Clause 1.2.7, "Reference standards: Group 7".

For class 4.8, 5.8, 6.8, 8.8 and 10.9 bolts, the design shear resistance $F_{v,Rd}$ should be taken as 0.85 times the value given in Table 4.8.

4.3.9.1 Design of slip-resistant shear connections

When the bolts are subjected to shear in the plane of the friction faces, the design slip resistance of a preloaded class 8.8 or 10.9 bolt should be calculated from the following equation:

$$\text{design slip resistance} = F_{s,Rd} = k_s n \mu F_{p,C} / \gamma_{M3} \quad (3.6)$$

where $k_s = 1.0$ for normal holes (for other types of holes, refer to Table 3.6 of Eurocode 3, Part 1-8), n is the number of friction surfaces, and μ is the slip factor, obtained either by specific tests on the friction surface in accordance with Clause 1.2.7 or, when relevant, as given in Table 4.9.

For class 8.8 and 10.9 bolts conforming with Clause 1.2.4, "Reference standards: Group 4", with controlled tightening in conformity with Clause 1.2.7, "Reference standards: Group 7", the preloading force $F_{p,C}$ should be calculated from the following equation:

$$\text{preloading force} = F_{p,C} = 0.7 f_{ub} A_s \quad (3.7)$$

Table 4.9. Slip factor μ for preloaded bolts (based on Table 3.7 of Eurocode 3, Part 1-8)

Class of friction surfaces (see Clause 1.2.7, "Reference standards: Group 7")	Preparation	Treatment	Slip factor μ
A	Blasted with shot or grit	Loose rust removed, no pitting. Spray metallized with aluminium. Spray metallized with a zinc-based coating that has been demonstrated to provide a slip factor of at least 0.5.	0.5
B	Blasted with shot or grit	Spray metallized with zinc	0.4
C	Wire brushed, flame cleaned	Loose dust removed, tight mill scale	0.3
D	Untreated, galvanized	Untreated	0.2

The slip factor μ is the ratio of the load per effective interface required to produce slip in a pure shear joint to the nominal shank tension. When the interface is shot blasted and spray metallized with aluminium after removal of rust and pitting, then $\mu = 0.5$.

But, in practice, the above rigorous procedure of surface preparation is difficult to follow at a work site. The surface is generally shot blasted and spray metallized with zinc instead. In the case of this type of surface preparation, a value of $\mu = 0.4$ is normally taken.

The tensile stress area A_s is given in Table 4.10 for various diameters of bolts.

For example, assume that the diameter of the HSFG bolts, class 8.8 (in a plate of grade S 275), is M24. For a bolt in double shear (double interface), $n = 2$. The slip factor (for a bolt blasted with shot, and no pitting, and spray metallized with aluminium) is

$\mu = 0.5$.

Coefficient for standard clearance hole $k_s = 1$.

Preloading force $F_{p,C} = 0.7 f_{ub} A_s = 0.7 \times 800 \times 353/10^3 = 198$ kN.

Design slip resistance $F_{s,Rd} = k_s n \mu F_{p,C}/\gamma_{M3} = 1 \times 2 \times 0.5 \times 198/1.25 = 158$ kN.

Table 4.11 gives the preloading force for HSFG bolts. The calculations of the preloading force (roof loads) are based on the net area of the bolt at the bottom of the threads.

4.3.9.2 Combined tension and shear

If a slip-resistant connection is subjected to an applied tensile force $F_{t,Ed}$ or $F_{t,Ed,ser}$, in addition to a shear force $F_{v,Ed}$ or $F_{v,Ed,ser}$ tending to produce slip, the design slip resistance per bolt should be taken from the following equations:

for a category B connection, $F_{s,Rd,ser} = k_s n \mu (F_{p,c} - 0.8 F_{t,Ed,ser})/\gamma_{M3,ser}$ (3.8a)

for a category C connection, $F_{s,Rd} = k_s n \mu (F_{p,c} - 0.8 F_{t,Ed})/\gamma_{M3}$ (3.8b)

If, in a moment connection, a contact force on the compression side counterbalances the applied tensile force, no reduction in slip resistance is required.

4.3.10 Design of bolts and splice plates in flanges and web of gantry girder

Consider a splice joint in a section 5.7 m from a support (a bearing).

Table 4.10. Diameter of bolt (d_0) and tensile stress area

Diameter of bolt d_0 (mm)	Tensile stress area A_s (mm^2)
12	84.3
16	157
20	245
22	303
24	353
27	459
30	561

4.3.10.1 Design considerations

- The connection should be designed to transfer moment, shear and thrust.
- The designed section of the gantry girder is a deep welded-plate girder.
- The bolts in the flanges are be designed to transfer the entire force in the flanges developed owing to the moment. The web does not take part in sharing any moment.
- It is assumed that the bolts in the web will transfer the full shear in the section considered.
- The thrust in the joint will be shared by bolts in both the flanges and the web.
- Since a gantry girder cannot be transported to the site if it is more than 12 m in length owing to highway regulations, we assume three pieces of girder separately fabricated at the fabrication shop, with a central piece 12 m long and two end pieces 6 m long. The pieces are spliced together on site and erected as one piece.

4.3.10.2 Calculations of moments and shears in a section 5.7 m from end support

Bending moment

Moment due to crane wheel loads. We draw an influence line diagram at 5.7 m from the end support (see Fig. 4.2). From the influence line diagram due to the crane wheel loads,

dynamic BM at 5.7 m = 728 × (4.31 + 4.02 + 3.65 + 3.36 + 1.83 + 1.53 + 1.17 + 0.88).

BM at 5.7 m = dynamic BM = 15 106 kN m.

With γ_{crk} = 1.5 (for moving loads due to cranes), the ultimate design moment at 5.7 m from the support is

$M_u = 1.5 \times 15\,106 = 22\,659$ kN m.

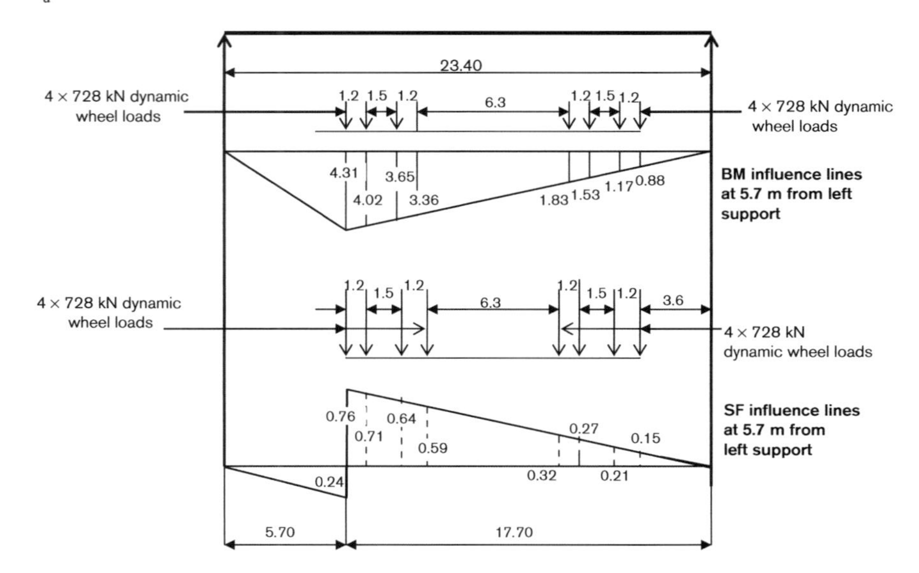

Figure 4.2 BM and shear force (SF) influence lines at 5.7 m from left support

Moment due to self-weight of crane. The assumed self-weight of the girder is 279 kN (previously calculated).

Self-weight/m length = 279/24 = 11.9 kN/m.

Reaction R_1 = 139.5 kN.

BM at 5.7 m from support $M_d = 139.5 \times 5.7 - 11.9 \times 5.7^2/2 = 602$ kN m.

With $\gamma_{Gk} = 1.35$ (for dead load),

ultimate design BM at 5.7 m = $1.35 \times 602 = 813$ kN m.

Therefore

total ultimate design BM at 5.7 m from support = M_u = 22 659 + 813 = 23 472 N m

and

total service BM at 5.7 m from support = (15 106 + 602) kN m = 15 708 kN m.

Shear

Shear due to crane wheel loads. We draw an influence line diagram at 5.7 m from the support (see Fig. 4.2). From the influence line diagram, the dynamic vertical shear at 5.7 m from the support is

$V = 728 \times (0.76 + 0.71 + 0.64 + 0.59 + 0.32 + 0.27 + 0.21 + 0.15) = 2657.2$ kN.

With $\gamma_{crk} = 1.5$, the ultimate design shear at 5.7 m from the support is

$V_u = 1.5 \times 2657.2 = 3986$ kN.

Shear due to self-weight $V_u = 1.35 \times (139.5 - 11.9 \times 5.7) = 96$ kN.

Total ultimate shear $V_u = 3986 + 96 = 4082$ kN

and

total service shear at 5.7 m from support $V_s = 2657.2 + 71.4 = 2729$ kN.

4.3.10.3 Design of number of bolts in flange (see Fig. 4.3)

Ultimate design moment at 5.7 m from support M_u = 22 659 kN m (calculated before). It is assumed that the whole moment will be taken up by the flanges.

Depth of plate girder D = 25 000 mm;

with 55 mm flange plates, the effective depth between the centres of gravity of the flange plates is 2500 − 55 = 2445 mm. Therefore

force in the flange = $F_t = F_c$ = 22 659/2.445 = 9267 kN

total horizontal shear force on the bolts = F_s = 9267 kN

Using M24 HSFG bolts (preloaded),

slip resistance/M24 bolt in double shear = $F_{s,Rd} = k_s n \mu F_{p,c}$

We assume $k_s = 1.0$ for a standard clearance hole, μ = slip factor = 0.5 (assuming the bolts are shot blasted and spray metallized with aluminium), $F_{p,c}$ = 198 kN (the preloading force (proof load) for class 8.8 M24 bolts with f_{ub} = 800 N/mm^2) (see Table 4.11) and n (for double shear) = 2. Therefore

$$F_{s,Rd} = 1 \times 2 \times 0.5 \times 198 = 198 \text{ kN}$$

Number of bolts required = 9267/198 = 46.8, say 52

Therefore we provide six rows of M24 HSFG bolts with ten in each row, total number = 60.

4.3.10.4 Design of number of bolts in the web (see Fig. 4.3)

Total ultimate shear in the section = $V_u = F_v$ = 4082 kN (already calculated above).

The total shear will be taken up by the bolts. We assume M24 HSFG bolts in double shear.

Slip resistance/bolt = $F_{s,Rd}$ = 198 kN (calculated above)

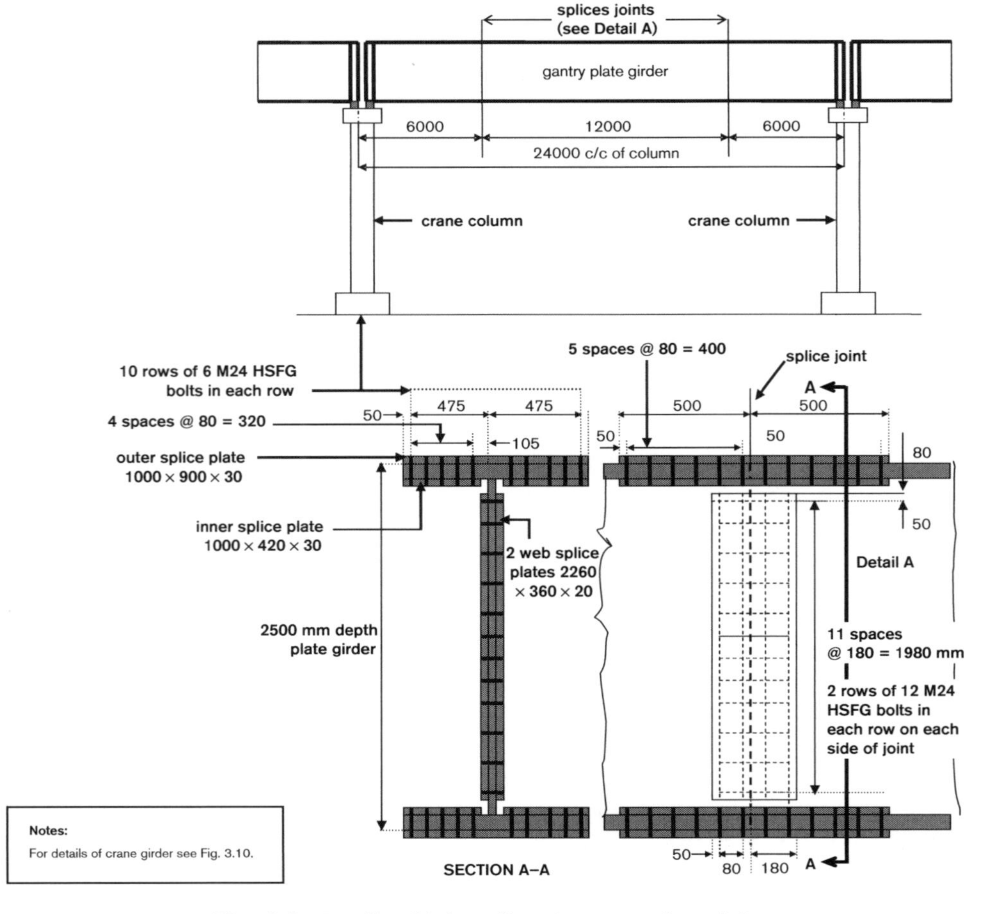

Fig. 4.3. Details of joint splices in gantry plate girder

Therefore

number of bolts required = 4082/198 = 21

We assume two rows of bolts consisting of 12 bolts in each row on each side of the splice joint.

Therefore we provide 24 M24 HSFG bolts in two rows.

4.3.10.5 Design of splice plates for the flanges

Total ultimate tension in the flange = F_t = 9267 kN

Area of splice plates required = $A_s = F_t/f_y$

For plates of steel grade S 275 and thickness less than 40 mm, f_y = 275 N/mm^2. Therefore the net area of splice plates required is

A_s = 9267 × 103/275 = 33 700 mm^2

Assume an outer 30 mm cover plate = 900 × 30 = 27 000 mm^2 and two inner 30 mm cover plates = 2 × 420 × 30 = 25 200 mm^2.

Total gross area provided = 27 000 + 25 200 = 52 200 mm^2

Deduct six 26 mm holes in a row:

area deducted = 10 × 26 × 55 = 14300 mm^2

Net area provided A_s = 52 200 − 14 300 = 379 000 mm^2 > A_t (33 700 mm^2 required).

Therefore we provide an outer splice plate 1000 × 900 × 30 mm and inner splice plates 2 × 1000 × 420 × 30 mm.

4.3.10.6 Design of splice plates for the web

Effective net area of web to resist shear = A_{web} = (2500 − 110) × 35 − 12 × 26 × 35
= 72 730 mm^2

The area of the splice plates should not be less than the area of the web = 72 730 mm^2. We assume two splice plates, net area = 2 × (2390 − 80 − 80) × 20 mm thick.

Area of splice plates = 2 × 2230 × 20 − 2 × 12 × 26 × 20 (deduction for 12 26 mm diameter holes)
= 76 720 mm^2 > 72 730 mm^2

Therefore we provide two web splice plates 2260 mm × 360 × 20 mm thick (see Fig. 4.3).

4.3.11 Design of bolts and splice plates in joint in column of stanchion A

Consider the splice connection in the crane column of stanchion A halfway between the base and the cap (see Fig. 4.4). Assume that all of the loads are transferred by the splice

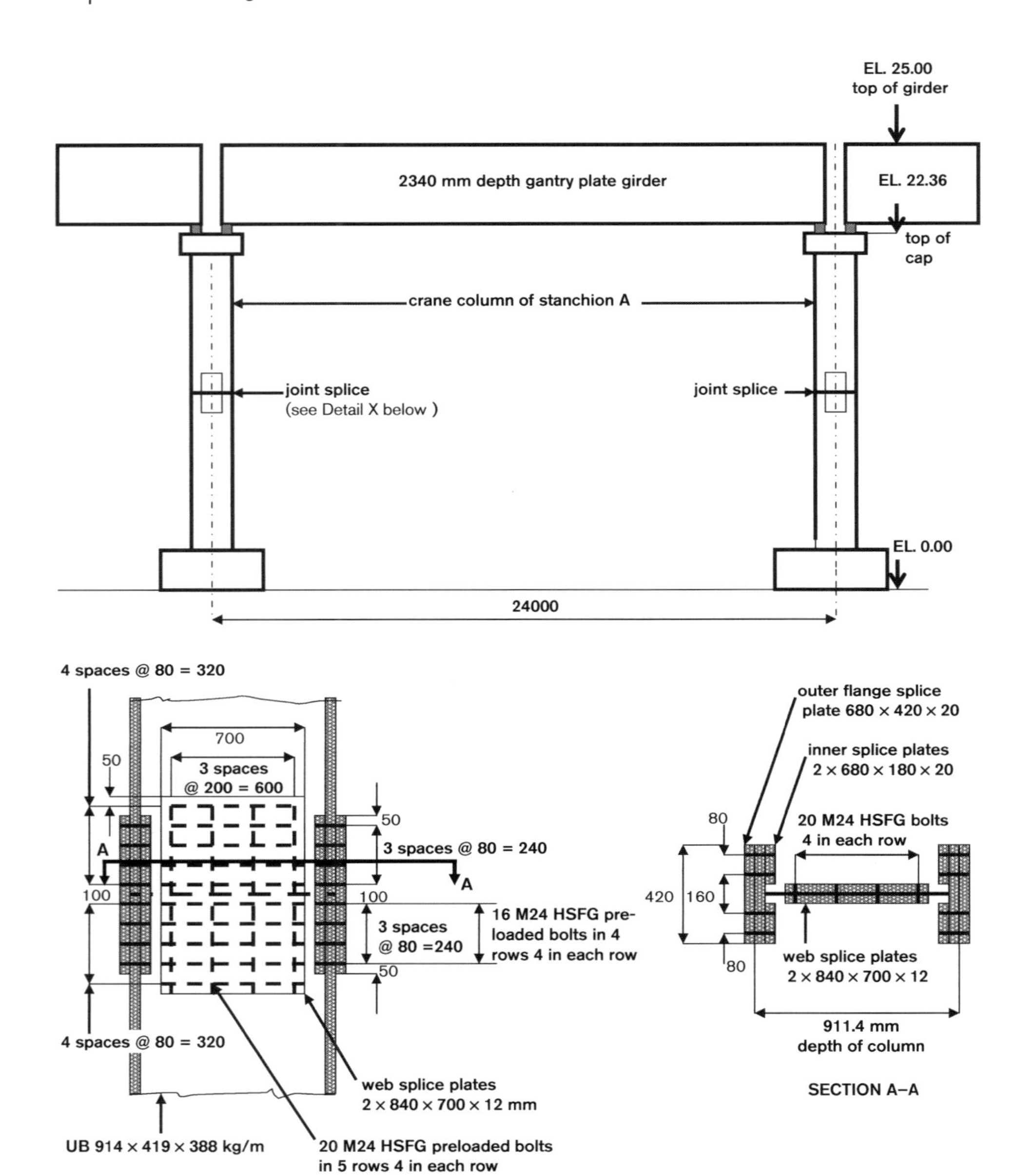

Fig. 4.4 Melting shop: details of splice in column

plates through HSFG preloaded bolts. The maximum ultimate load (thrust) in the crane column is

$$N_{\mathrm{Ed,max}} = 9970 \text{ kN (previously calculated)}$$

The section of the member, already designed, is UB914 × 419 × 388 kg/m; $A_g = 493.9$ cm^2. The total load will be shared by the flanges and the web in proportion to their areas.

Area of flanges $A_f = 420.5 \times 36.6 \times 2 = 30\,781\ \text{mm}^2$.

Area of web $A_w = 49\,390 - 30\,781 = 18\,609\ \text{mm}^2$.

Therefore

load shared by flanges = 9970 × 30 781/49 390 = 6214 kN

load shared by web = (9970 − 6214) = 3756 kN

4.3.12 Design of bolt connection in the flange

From the above calculations,

total load carried by each flange = $N_{Ed,f}/2 = 6214/2 = 3107$ kN

Assume M24 HSFG preloaded bolts.

Slip resistance/bolt (with $f_{ub} = 800\ \text{N/mm}^2$) = $F_{s,Rd} = 198$ kN (in double shear).

Therefore

number of bolts required = 3107/198 = 15.7

Therefore we provide 16 M24 HSFG preloaded bolts in flanges, four in each row.

4.3.13 Design of bolt connection in the web

From the above calculations,

total load carried by web = 3756 kN

Assume M24 HSFG preloaded bolts.

Slip resistance/bolt = 198 kN

Therefore

number of bolts required = 3756/198 = 19

Therefore we provide 20 M24 HSFG preloaded bolts in web, four in each row.

4.3.14 Design of splice plates in flanges

From the above calculations,

total load carried by each flange = 3107 kN

Thickness of flange = $t_f = 36.6$ mm

With the design strength of steel grade S 275, $f_y = 275\ \text{N/mm}^2$, the permissible strength f_y for 2 × 20 mm splice plates is 275 N/mm².

Area of splice plates required $A_f = 3107 \times 10^3/275 = 11\,298\ \text{mm}^2$

We assume that the area of the outer splice plate is $420 \times 20 = 8400$ mm^2 and the area of the inner splice plates is $2 \times 165 \times 20 = 6600$ mm^2.

Total area of splice plates = 8400 + 6600 = 1500 mm^2 > 11 298 mm^2

Therefore we provide an outer splice plate 420 × 20 mm and two inner splice plates 180 × 20 mm.

4.3.15 Design of splice plates for web

From the above calculations,

total load carried by web = 3756 kN

Thickness of web = t = 21.4 mm

Area of splice plates required $A_{web} = 375 \times 10^3/275 = 13\ 658$ mm^2

Using 2 × 700 × 12 splice plates

area = 16 800 mm^2 > 13 658 mm^2

Therefore we provide two 700 × 12 × 840 mm long web splice plates (see Fig. 4.4).

References

British Standards Institution, 1998. BS EN ISO 14555: 1998, Welding. Arc stud welding of metallic materials.

British Standards Institution, 1999. BS EN 12345: 1999, Welding. Multilingual terms for welded joints with illustrations.

British Standards Institution, 2003. BS EN 13918: 2003, Gas welding equipment. Integrated flowmeter regulators used on cylinders for welding, cutting and allied processes. Classification, specification and tests.

British Standards Institution, 2008. BS EN 1090-2: 2008, Execution of steel structures and aluminium structures. Technical requirements for the execution of steel structures

Eurocode, 2005a. BS EN 1993-1-8: 2005, Eurocode 3. General. Design of joints.

Eurocode, 2005b. BS EN 1993-1-1: 2005, Eurocode 3. Design of steel structures. General rules and rules for buildings.

Eurocode, 2005c. BS EN 1993-1-9: 2005, Eurocode 3. Design of steel structures. Fatigue.

CHAPTER 5

Design of Purlins, Side Rails, Roof Trusses, Roof Girders, Intermediate Columns and Horizontal Roof Bracings

5.1 Purlins in melting bay (members subjected to bending)

5.1.1 Method of design

The purlins and side rails may be designed on the assumption that the cladding provides lateral restraint to the compression flange, provided the cladding and fixings act with adequate capacity to restrain the members. The purlins and side rails may be designed by either an empirical method or an analytical method (treating the members as normal structural beams).

5.1.1.1 Empirical method

The purlins and side rails may be designed by an empirical method by applying empirical formulae, provided the following conditions are satisfied:

- The effective span should not exceed 6.5 m from centre to centre of the supports.
- The members should be of a minimum steel grade of S 275, with yield strength $f_y = 275$ N/mm^2 and with a thickness $t \leq 40$ mm.
- There is no requirement to adopt factored loads in calculating the section of the members. So, unfactored loads should be used in the empirical design.
- If the members are simply supported, at least two connecting bolts should be provided.
- If the members are continuous over two or more supports, the joints should be staggered over the adjacent supports.
- The roof slope should not exceed 30°.

In our case, the maximum spacing of the roof trusses is assumed to be 6.0 m, and grade S 275 steel is to be used. Purlins of 6 m span will result an economical size of the section. A span in excess of 6.5 m would involve lengthy structural calculations and we would not arrive at a suitable economical section of the purlin.

So, we may use an empirical method. But Eurocode 3 does not specify the adoption of an empirical method.

5.1.1.2 Analytical method

So, in our case we shall adopt an analytical method and follow Eurocode 3, Part 1-1. (All of the references to clauses and equations in this chapter relate to this Eurocode unless otherwise stated.)

5.1.2 Design data

Span = 6.0 m, continuous over roof trusses

Spacing = 2.25 m

(This 2.25 m spacing of the purlins has been adopted because the spacing of the node points of the top chord of the roof truss is 2.25 m to avoid the development of any local bending moment in the top chord due to eccentricity of loads.)

5.1.3 Loadings

These are calculated based on Eurocode 1, Part 1-1 (Eurocode, 2002a)

5.1.3.1 Dead loads

We assume 20G corrugated galvanized steel sheeting, including side laps two corrugations wide, and 150 mm end laps.

Load due to sheeting = 0.11 kN/m^2

Insulation (10 mm fibre board) = 28 N/m^2 = 0.028 kN/m^2

Total = 0.138 kN/m^2

Load/m of span (assuming 2.25 m spacing of purlins) = 0.138 × 2.25 = 0.31 kN/m

Self-weight of purlin (assuming a channel 150 × 89 × 23.84 kg/m) = 0.24 kN/m

Therefore

total dead load/m = (0.31 + 0.24) kN/m = g_k = 0.55 kN/m

total dead load = G_k = 0.55×6 = 3.3 kN

5.1.3.2 Live loads

For a roof slope not greater than 10°, with access, the imposed load q_k is 1.5 kN/m^2. In our case,

roof slope = tan θ = 1.5/13.5 = 0.11, θ = 6.28°

So, q_k = 1.5 kN/m^2

Imposed load/m = q_k = 1.5 × 2.25 = 3.38 kN/m; Q_k = 3.38 × 6 = 20.3 kN

5.1.3.3 Snow loads

Not considered.

5.1.3.4 Wind loads

These are calculated based on Eurocode 1, Part 1-4 (Eurocode, 2005a).

Wind velocity pressure (dynamic) = $q_p(z)$ = 1.27 kN/m^2

(previously calculated, see Chapter 2). When the wind is blowing from right to left, the resultant wind pressure coefficient on a windward slope with internal suction is

c_{pe} = −0.30 upwards

(previously calculated, see Chapter 2), and, also, the resultant pressure coefficient on a windward slope with positive internal pressure is

$c_{pe} = -0.90$ upwards

(previously calculated, see Chapter 2). Therefore the external wind pressure normal to the roof is

$p_e = q_p c_{pe} = 1.27 \times -0.90 = -1.14\ \text{kN/m}^2$

The vertical component of the wind pressure is

$p_{ev} = p_e \cos\theta = -1.14 \times \cos 6.28° = 1.14 \times 0.99 = -1.13\ \text{kN/m}^2$ acting upwards.

Therefore the total wind force on the whole span (6.0 m) with 2.25 m spacing is

$W_k = -1.13 \times 2.25 \times 6 = -15.3$ kN acting upwards

This is greater than the total dead load (3.3 kN). So, there is the possibility of reversal of stresses in the purlin.

5.1.4 Moments

We assume that the purlin is continuous over the roof trusses. Since the length of a member is limited to 12 m owing to transport regulations, the purlin consists of two spans.

5.1.4.1 Characteristic moments

If we consider characteristic moments over the support, we have the following values.

Due to dead load (DL): $M_{gk} = 0.125 G_k L = 0.125 \times 3.3 \times 6 = 2.5$ kN m, load acting downwards.

Due to live load (LL): $M_{qk} = 0.125 Q_k L = 0.125 \times 20.3 \times 6 = 15.2$ kN m, load acting downwards.

Due to wind load (WL): $M_{wk} = 0.125 W_k L = 0.125 \times 15.3 \times 6 = 11.9$ kN m, load acting upwards.

If we consider moments at the midpoint of the end span, we have the following values.

Due to DL: $M_{gk} = 0.07 G_k L = 0.07 \times 3.3 \times 6 = 1.4$ kN m, load acting downwards.

Due to LL: $M_{qk} = 0.096 Q_k L = 0.096\ 20.3\ 6 = 11.7$ kN m, load acting downwards.

Due to WL: $M_{wk} = 0.096 W_k L = 0.096 \times 15.3 \times 6 = 8.8$ kN m, load acting upwards.

(The moment coefficients have been taken from the *Reinforced Concrete Designer's Handbook* (Reynolds and Steedman, 1995).)

5.1.4.2 Ultimate design moments for load combinations

We calculate the following partial safety factors for load combinations in the ULS design method, with reference to Table A1.2 (B) of BS EN 1990: 2002(E) (Eurocode, 2002b).

Case 1: when dead load and live load are acting together, no wind.

Partial factor for permanent actions (DL) = γ_{Gj} = 1.35 (unfavourable).

Partial factor for leading variable action (LL) = γ_{Qk1} = 1.5.

Therefore

ultimate design moment at *support* = M_u

$= \gamma_{Gj}G_k + \gamma Q_k Q_k = 1.35 \times 2.5 + 1.5 \times 15.2 = 23.4$ kN m;

ultimate design moment at *midspan* = $M_u = 1.35 \times 1.4 + 1.5 \times 11.7 = 19.4$ kN m.

Case 2: when dead load + live load + wind load are acting together.

Partial factor for permanent actions (DL) = γ_{Gj} = 1.35 (unfavourable).

Partial factor for leading variable action (LL) = γ_{Qk1} = 1.5.

Partial factor for accompanying variable actions (WL) = $\gamma_{wk1}\psi_0 = 1.5 \times 0.6 = 0.9$

(for the value of ψ, refer to Table A1.1 of BS EN 1990: 2002(E)). Therefore

ultimate design moment at support = $M_u = \gamma G_j G_k + \gamma Q_k Q_k + \gamma W_k \psi_0 W_k$

$= 1.35 \times 2.5 + 1.5 \times 15.2 - 1.5 \times 0.6 \times 11.9 = 12.7$ kN m;

ultimate design moment at midspan = M_u

$= 1.35 \times 1.4 + 1.5 \times 11.7 - 0.9 \times 8.8 = 11.5$ kN m.

Case 3: when minimum dead load + wind load but no live load.

Partial factor for permanent actions (DL) = γ_{Gj} = 1.0 (favourable)

Partial factor for leading variable wind action (WL) = γ_{Wk} = 1.5.

Therefore

ultimate design moment at support = $M_u = \gamma_{Gk}G_k - \gamma_{Wk}W_k$

$= 1.0 \times 2.5 - 1.5 \times 15.2 = -20.3$ kN m (acting upwards);

ultimate design moment at midspan = $M_u = 1.0 \times 1.4 - 1.5 \times 8.8 = -11.8$ kN m.

Therefore, case 1 gives the maximum ultimate design moment at the support, equal to 23.4 kN m, and the maximum ultimate design moment at midspan, equal to 19.4 kN m.

5.1.4.3 Ultimate shear at support

Ultimate shear at central support = $V_{ed} = (1.35 \times 3.3 + 1.5 \times 20.3) \times 0.626 = 21.8$ kN.

Ultimate shear at end support = $V_{ed} = (1.35 \times 3.3 + 1.5 \times 20.3) \times 0.375 = 13.1$ kN.

5.1.5 Design of section

The section will be designed in accordance with Eurocode 3, Part 1-1 (Eurocode, 2005b).

5.1.5.1 Design strength

The design yield strength f_y is that of grade S 275 steel: $f_y = 275$ N/mm^2 ($t \le 40$ mm).

5.1.5.2 Initial sizing of section

In an elastic analysis, to limit the deflection of the purlin to $L/200$, the minimum depth of the member should be as follows:

- For a simply supported member: depth = $h = L/20$, where L is the span.
- For a continuous member: depth = $h = L/25$.

In our case, $L = 6000$ mm, and the purlin is assumed to be continuous. So,

minimum depth of purlin = 6000/25 = 240 mm.

Try a section UB305 × 165 × 40 kg/m (see Fig. 5.1).

Depth of section = h = 303.4 mm.

Depth of straight portion of web = d = 265.2 mm.

Width of flange = b = 116.5 mm.

Thickness of flange = t_f = 10.2 mm.

Thickness of web = t_w = 6.0 mm.

Root thickness = r = 8.9 mm.

Radius of gyration about major axis = r_y = 12.9 cm.

Radius of gyration about minor axis = r_z = 3.96 cm.

Plastic modulus about major axis = W_y = 623 cm^3.

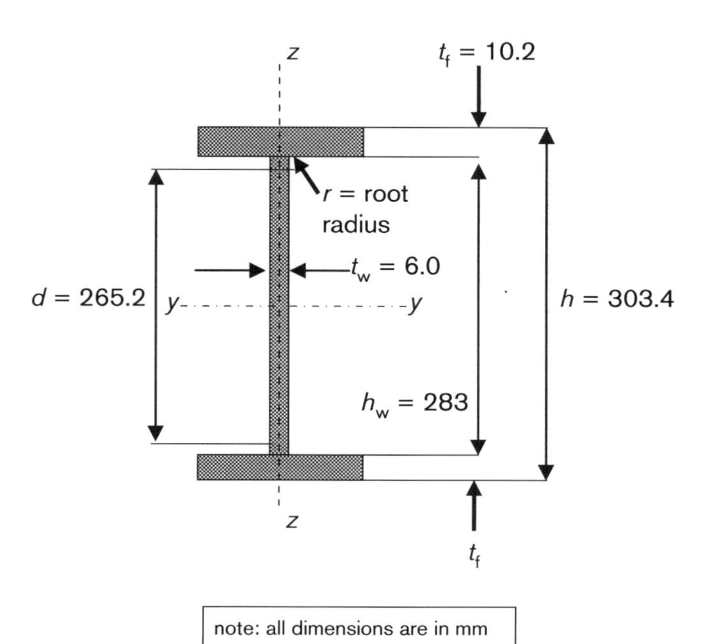

Fig. 5.1. Purlin section, UB305 × 165 × 40 kg/m (all dimensions in mm)

Elastic modulus about major axis = Z_y = 560 cm^3.

Cross-sectional area = A = 51.3 cm^2.

5.1.5.3 Section classification

Flange

Stress factor = $\varepsilon = \sqrt{235/f_y} = \sqrt{235/275} = 0.92$.

Outstand of flange = $c = [b - (t_w + 2r)]/2 = [165 - (6 + 2 \times 8.9)]/2 = 70.4$ mm.

Ratio $c/t_f = 70.4/10.2 = 6.9$.

Referring to Table 5.2 (sheet 1) of Eurocode 3, Part 1-1 (see Appendix B), for class 1 section classification, the limiting value of $c/t_f \leq 9\varepsilon$ ($9 \times 0.92 = 8.28$), i.e. $6.9 < 8.28$. So, the flange satisfies the conditions for class 1 classification.

Web

Ratio $d/t_w = 265.2/6 = 44.2$.

$72\varepsilon = 72 \times 0.92 = 66.2$.

Referring to Table 5.2 (sheet 2) of Eurocode 3, Part 1-1, for class 1 section classification, the limiting value of $d/t_w \leq 72\varepsilon$, i.e. $44.2 < 66.2$. So, the web satisfies the conditions for class 1 classification.

5.1.5.4 Moment capacity

Maximum ultimate design moment over support = $M_u = M_{y,Ed}$ = 23.4 kN m

(load case 1: resultant load acting downwards, already calculated).

Maximum ultimate design moment at midspan = $M_{y,Ed} = -11.8$ kN m

(load case 3: resultant load acting upwards, already calculated).

The moment capacity should be calculated in the following ways:

- *When the web is not susceptible to buckling*. When the web depth-to-thickness ratio $d/t_w \leq 72\varepsilon$ for class 1 classification, it should be the case that the web is not susceptible to buckling, and the moment capacity should be calculated from the equation $M_{y,Rd} = f_y S_y$ provided the shear force in the section satisfies the condition $V_{ed} < 0.5V_{pl,Rd}$. In our case, we have assumed that d/t_f (44.2) < 72ε (66.2). So, the web is not susceptible to buckling.
- *When the ultimate shear force $V_{ed} \leq 0.5V_{pl,Rd}$.*

Ultimate shear force at penultimate support = V_{ed} = 21.8 kN (calculated before), and, referring to equation (6.18) of Eurocode 3, Part 1-1,

design plastic shear capacity of section = $V_{pl,rd} = A_v(f_y/\sqrt{3})/\gamma_{Mo}$,

where $A_v = A - 2bt_f + 2(t_w + 2r)t_f = 2250$ mm^2.

But A_v should not be less than $h_w t_w$, i.e. $265 \times 6 = 1591$ mm^2 < 2250 mm^2. So, $A_v = 2250$ mm^2 and $f_y/\sqrt{3} = 158.8$ N/mm^2. And, referring to Clause 6.1 of Eurocode 3, Part 1-1, $\gamma_{Mo} = 1$. Therefore,

design plastic shear capacity $V_{pl,Rd} = 2250 \times 158.8/10^3 = 357$ kN.

In our case, $V_{ed} < 0.5 \times V_{pl,rd}$, i.e. 21.8 kN < 0.5 × 357 kN, which satisfies the condition. So, no reduction of moment capacity needs to be considered. Therefore

plastic moment capacity $M_{ypl,Rd} = f_y S_y = 275 \times 623 \times 10^3/10^6$

$= 171$ kN m >> M_{yEd} (23.4 kN m). OK

5.1.5.5 Shear buckling resistance

The shear buckling resistance need not be checked if the ratio $d/t_w \leq 72\varepsilon$. In our case, $d/t_w = 265.2/6 = 44.2$ and $72\varepsilon = 72 \times 0.92 = 66.2$. So,

$$d/t_w < 72\varepsilon$$

Therefore, the shear buckling resistance need not be checked.

5.1.5.6 Buckling resistance moment

It has been explained earlier, in Chapter 3, that if a structural member is subjected to bending owing to external load actions, the unrestrained length of the compression flange moves laterally at right angles to its normal direction of bending, thus creating a lateral–torsional buckling of the member as shown in Fig. 3.3. So, we have to examine if the resisting moment due to buckling (the buckling resistance moment) is greater than the bending moment in the member due to the loadings.

In our case, the flanges of the structural member will be subjected to reversal of stresses owing to the loading conditions described below and shown in Fig. 5.2.

In *case 1*, when the dead and live loads without wind loads act together because the bending moments act downwards, the compression of the top flange due to the bending moment at midspan is restrained by the roof cladding (which is assumed to be rigid enough to resist buckling). But the bottom flange undergoes compressive stress over the support. We may consider the point of contraflexure in the bending moment diagram (see Fig. 5.2(a)) as a point restraint, which is assumed to occur at a distance of about one-quarter of the span from either side of the support.

In *case 3*, when the minimum dead load and the wind load act simultaneously without live loads, the combined-ultimate-moment diagram in Fig. 5.2(c) shows that the bottom flange is under compression. Considering the point of contraflexure in the bending moment diagram as a point of restraint of the compression flange, the length of the unrestrained compression flange may be taken as equal to three-quarters of the span.

As a result, in case 1, at the support, the compression flange is subject to lateral–torsional buckling for a length of about one-quarter (assumed) of the span on either side of the support, thus reducing the resisting moment. In case 3, at midspan, three-quarters of the span of the bottom flange of the purlin, considered as a continuous member, is subject to lateral–torsional buckling, thus again reducing the resisting moment. So, we have to investigate both of these cases.

The buckling resistance moment may be calculated in the following way. Firstly, consider case 1. The length between the restraints of the bottom compression flange is $L_c = 6000/4 = 1500$ mm, and $M_u = 23.4$ kN m. Referring to the equation given in Clause 6.3.2.1, the design buckling resistance moment is

$$M_{b,Rd} = \chi_{LT} W_y f_y / \gamma_{M1} \tag{6.55}$$

where χ_{LT} is the reduction factor to be calculated, W_y is the section modulus along the major axis = 623 cm^3, f_y is the ultimate strength = 275 N/mm^2 and γ_{M1} is the partial factor

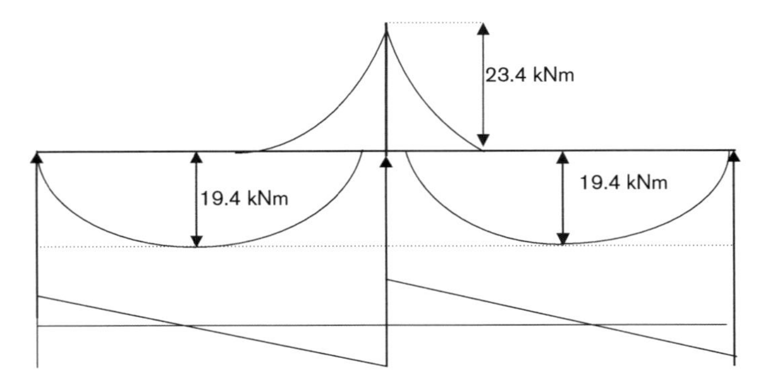

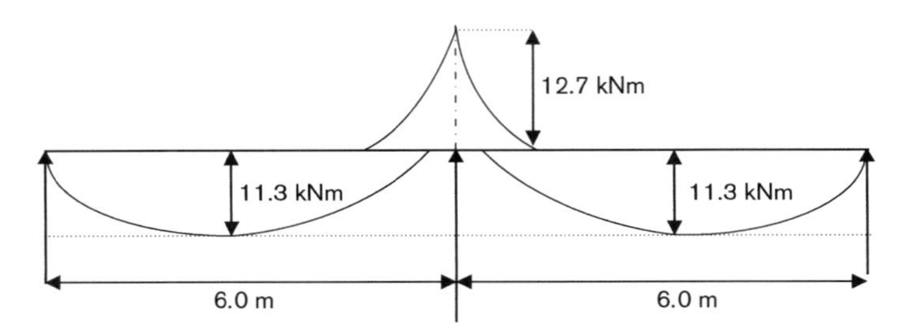

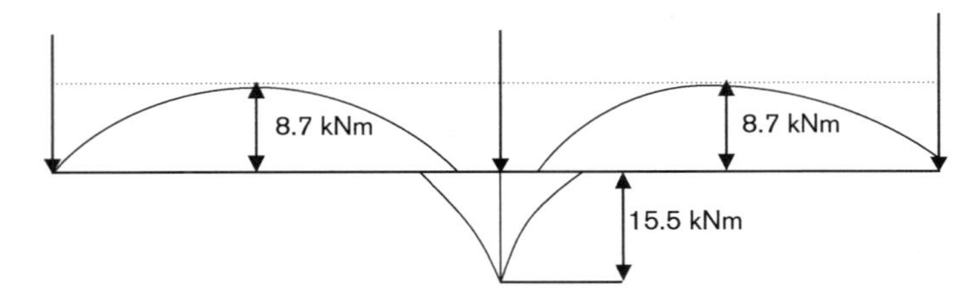

Fig. 5.2. Ultimate design moment and shear under various loading conditions

for resistance of member to instability = 1 (see Clause 6.1). For rolled or equivalent welded sections in bending, values of χ_{LT} may be determined from the following equation:

$$\chi_{LT} = 1/[\Phi_{LT} + \sqrt{(\Phi_{LT}^2 - \beta\bar{\lambda}_{LT}^2)}] \quad \text{but} \quad \chi_{LT} \leq 1 \quad \text{and} \quad \chi_{LT} \leq 1/\bar{\lambda}_{LT}^2 \qquad (6.57)$$

where $\Phi_{LT} = 0.5[1 + \alpha_{LT}(\bar{\lambda}_{LT} - \bar{\lambda}_{LT,0}) + \beta\bar{\lambda}_{LT}^2]$; $\bar{\lambda}_{LT,0}$ is a parameter, recommended value for rolled sections = 0.4 (maximum); β is a parameter, recommended value = 0.75 (minimum); and

$$\bar{\lambda}_{LT} = \sqrt{W_y f_y / M_{cr}}$$

where M_{cr} is the elastic critical moment for critical lateral buckling. The code does not specify how to calculate this value. So, we shall adopt a simplified approach.

Referring to Clause 6.3.2.4, "simplified assessment methods for beams with restraints in building", of Eurocode 3, Part 1-1, the compression flange will not be susceptible to lateral buckling if the length L_c between restraints or the resulting slenderness $\bar{\lambda}_f$ of the equivalent compression flange satisfies the following condition:

$$\bar{\lambda}_f = k_c L_c/(i_{f,z}\lambda_1) \leq \lambda_{c0}(M_{c,Rd}/M_{y,Ed}) \quad (6.59)$$

where $M_{y,Ed}$ is the maximum design value of the bending moment within the restraint spacing, equal to 23.4 kN m; $M_{c,Rd}$ is the design resistance for bending about the y–y axis of the cross-section, equal to $W_y f_y/\gamma_{M1} = 623 \times 10^3 \times 275/1/10^6 = 171$ kN m; k_c is the slenderness correction factor for the moment distribution between restraints, equal to 0.9 (see Table 6, "Correction factors", of Eurocode 3, Part 1-1); $i_{f,z}$ is the radius of gyration of the equivalent compression flange about the minor axis of the section, composed of the compression flange plus 1/3 of the compressed part of the web area, equal to 3.86 cm (1/3 of compression web neglected); and $\bar{\lambda}_{c,0}$ is the slenderness limit of this equivalent compression flange, recommended value $\bar{\lambda}_{LT,0} + 0.10$.

Referring to Clause 6.3.2.3 of Eurocode 3, Part 1-1, $\bar{\lambda}_{LT,0} = 0.4$ is recommended for I or welded sections, and so

$$\bar{\lambda}_{LT,0} = 0.4 + 0.1 = 0.5;$$

$$\lambda_1 = 93.9\varepsilon = 93.9 \times 0.92 = 86.4, \text{ where } \varepsilon = \sqrt{(235/f_y)} = \sqrt{(235/275)} = 0.92$$

$$L_c = \text{length between restraints} = 6000/4 = 150 \text{ cm}$$

Therefore

$$\bar{\lambda}_f = k_c L_c/(i_{f,z}\lambda_1) = 0.9 \times 154/(3.86 \times 86.4) = 0.40 \quad \text{and}$$
$$\bar{\lambda}_{c,0}(M_{c,Rd}/M_{y,Ed}) = 0.5 \times (171/23.4) = 3.65$$

So,

$$\bar{\lambda}_f\ (0.4) \leq \bar{\lambda}_{c,0}(M_{c,Rd}/M_{y,Ed})$$

Therefore, the compression flange is not susceptible to lateral–torsional buckling and no reduction of bending resistance due to lateral buckling needs to be considered. Therefore

design buckling resistance moment $M_{b,Rd} = M_{c,Rd} = 171 \text{ kN m} > M_{Ed}$ (23.4 kN m)

Next, consider case 3, when the wind force is predominantly acting upwards. The length between the restraint points of the bottom flange in the span is $0.75 \times 6000 = 450$ mm.

Moment at midspan = −11.8 kN m (resultant load acting upwards),

$$\bar{\lambda}_f = k_c L_c/(i_{f,z}\lambda_1) = 0.91 \times 450/(3.86 \times 86.4) = 1.22$$

and $\bar{\lambda}_{c,0}(M_{c,Rd}/M_{y,Ed}) = 0.5 \times (171/23.4) = 3.65 > 1.22$

$\times$ design resisting moment $= M_{b,Rd} = W_{pl,y} \times f_y = 623 \times 275/10^3$
$= 171 \text{ kN m} > 23.4 \text{ kN m}$ <u>OK</u>

Therefore, we adopt UB 305 × 165 × 40 kg/m for the purlin.

5.2 Side sheeting rails (members subjected to biaxial bending)

5.2.1 Method of design

We refer to the design of the purlins. The analytical method will be followed.

5.2.2 Design considerations

The side rails should be continuous over two spans of 6 m, as the restrictions on transport do not allow lengths of more than 12 m. The bending of the side rails due to vertical dead loads occurs in the minor axis of the member, and the bending due to wind loads occurs in the major axis.

5.2.3 Design data

Vertical spacing = 2 m.

Span = 6 m continuous over two spans.

5.2.4 Loadings

5.2.4.1 Dead loads

As calculated for purlins (assuming UB305 × 165 × 40 kg/m) = 0.154 kN/m.

Total dead load = $G_k = 0.154 \times 6 = 3.08$ kN.

5.2.4.2 Wind loads

Effective external pressure on vertical wall face = $p_e = q_p c_{pe} = 1.27 \times 1.1 = 1.4$ kN/m^2,

where q_p is the peak velocity pressure and c_{pe} is the resultant pressure coefficient, both already calculated.

Spacing of side rail = 2 m.

So, wind pressure/m run = 1.4 × 2 = 2.8 kN/m

× total wind pressure on side rail = $W_k = 2.8 \times 6 = 16.8$ kN.

5.2.5 Characteristic moments

5.2.5.1 Due to vertical dead loads (moments along minor axis)

Over the central support = $M_{gk,v} = 0.125 G_k L = 0.125 \times 3.08 \times 6 = 2.31$ kN m.

At midspan = $M_{gk,y} = 0.07 G_k L = 0.07 \times 3.08 \times 6 = 1.3$ kN m.

5.2.5.2 Due to horizontal wind loads (moments along major axis)

Over the central support = $M_{wk,z} = 0.125 W_k L = 0.125 \times 16.8 \times 6 = 12.6$ kN m.

At midspan = $M_{wk,z} = 0.096 W_k L = 0.096 \times 16.8 \times 6 = 9.7$ kN m.

5.2.6 Ultimate design moments

5.2.6.1 Due to vertical dead loads

The ultimate design moments are calculated with a safety factor for permanent actions (dead loads) $\gamma_{Gj} = 1.35$.

Ultimate design moment over the central support $= M_{ugk,y}$
$= \gamma_{Gj} M_{gk,y} = 1.35 \times 2.31 = 3.1$ kN m.

At midspan $= M_{ugk,y} = \gamma_{Gj} M_{gk,y} = 1.35 \times 1.3 = 1.8$ kN m.

5.2.6.2 Due to horizontal wind loads

The ultimate design moments are calculated with a safety factor for variable actions $\gamma_{wk} = 1.5$.

Ultimate design moment over the central support $= M_{uwk,z} = 1.5 M_{wk,z} = 1.5 \times 12.6$
$= 18.9$ kN m.

At midspan $= M_{uwk,z} = 1.5 M_{wk,z} = 1.5 \times 9.7 = 14.6$ kN m.

5.2.7 Ultimate shear at support

This is due to horizontal wind loads.

Ultimate shear at central support $= V_{ed} = 1.5 \times 16.8 \times 0.625 = 15.8$ kN m.

5.2.8 Design of section

The member will be designed in accordance with Eurocode 3, Part 1-1.

5.2.8.1 Design strength

The design yield strength f_y is that of grade S 275 steel, $f_y = 275$ N/mm^2.

5.2.8.2 Initial sizing of section

In an elastic analysis, to limit the deflection of the side rail to $L/200$, the minimum depth of the member should be as follows:

- For a simply supported member: depth $= h = L/20$, where L is the span.
- For a continuous member: depth $= h = L/25$.

In our case, $L = 6000$ mm, and the side rail is assumed to be continuous. So,

minimum depth of side rail = 6000/25 = 240 mm.

Try a section UB 305 × 165 × 40 kg/m (see Fig. 5.1).

Depth of section $= h = 303.4$ mm.

Depth of straight portion of web $= d = 265.2$ mm.

Width of flange $= b = 165$ mm.

Thickness of flange $= t_f = 10.2$ mm.

Thickness of web $= t_w = 6.0$ mm.

Root thickness $= r = 8.9$ mm.

Radius of gyration about major axis $= i_y = 12.9$ cm.

Radius of gyration about minor axis $= i_z = 3.96$ cm.

Plastic modulus about major axis $= W_{pl,y} = 623$ cm^3.

Plastic modulus about minor axis = $W_{pl,z}$ = 142 cm^3.

Elastic modulus about major axis = W_y = 560 cm^3.

Elastic modulus about minor axis = W_z = 92.6 cm^3.

Cross-sectional area = A = 51.3 cm^2.

5.2.8.3 Section classification

Flange

Stress factor $\varepsilon = \sqrt{(235/275)} = 0.92$.

Outstand of flange $c = (b - t_w - 2r)/2 = (165 - 6.0 - 2 \times 8.9)/2 = 70.4$ mm.

Ratio $c/t_f = 70.4/10.2 = 6.9$; $9\varepsilon = 8.28$.

Referring to Table 5.2 (sheet 1) of Eurocode 3, Part 1-1, the limiting value $c/t_f \leq 9\varepsilon$. So, the flange satisfies the conditions for class 1 classification.

Web

Ratio $d/t_w = 44.2$.

$72\varepsilon = 66.2$.

Referring to Table 5.2 (sheet 2) of Eurocode 3, Part 1-1, the limiting value $d/t_w \leq 72\varepsilon$. So, the web satisfies the conditions for class 1 classification.

5.2.8.4 Moment capacity

We consider the moments due to wind.

Maximum horizontal design moment along major axis over support = M_{uy}
= 18.9 kN m.

Maximum ultimate shear = V_{ed} = 15.8 kN

(the above values have already been calculated).

The moment capacity should be calculated in the following ways:

- *When the web is not susceptible to buckling.* When the web depth-to-thickness ratio = $d/t_w \leq 72\varepsilon$ for class 1 classification, it should be assumed that the web is not susceptible to buckling, and the moment capacity should be calculated from the equation $M_{yrd} = f_y W_{pl,y}$, provided that the shear force $V_{ed} < 0.5V_{pl,Rd}$. In our case, we have assumed a section in which $d/t_w < 72\varepsilon$. So, the web is not susceptible to buckling.
- *When the ultimate shear force $V_{ed} \leq 0.5V_{pl,Rd}$.* In our case,

maximum ultimate shear at central support V_{ed} = 15.8 kN,

and, referring to equation (6.18) of Eurocode 3, Part 1-1,

design plastic shear capacity of section in major axis = $V_{pl,Rd} = A_v f_y/\sqrt{3}y/\gamma_{M0}$,

where

$$A_v = A - 2bt_f + 2(t_w + 2r)t_f = 5130 - 2 \times 116.5 \times 10.2 + 2 \times (6 + 2 \times 8.9) \times 10.2 = 3239 \text{ mm}^2.$$

But A_v should not be less than $h_w t_w = (303.4 - 2 \times 10.2) \times 6 = 1698\ \text{mm}^2 < A_v$. OK
Therefore

$$V_{pl,Rd} = (3239 \times 275/(3)^{0.5})/10^3 = 514\ \text{kN}$$

Thus, $V_{ed} < 0.5 \times 514$ satisfies the condition. So, no reduction of moment capacity needs to be considered. Therefore

$$\text{plastic moment capacity } M_{yRd} = f_y W_{pl,y} = 275 \times 623 \times 10^3/10^6$$
$$= 171\ \text{kN m} < M_{yEd}\ (M_{uy} = 18.9\ \text{kN m}).$$ OK

5.2.8.5 Shear buckling resistance

The shear buckling resistance need not be checked if the ratio $d/t_w \leq 72\varepsilon$. In our case, $d/t_w = 265.2/6 = 44.2$ and $72\varepsilon = 66.2$. So,

$$d/t_w < 72\varepsilon$$

Therefore, the shear buckling resistance need not be checked.

5.2.8.6 Buckling resistance moment

Along the *major axis* of the member, when a horizontal wind pressure is acting,

$$M_{uy} = 18.9\ \text{kN m}$$

As explained before in the purlin calculations, the compression flange is connected to side sheeting at midspan and is restrained against torsional buckling. But the compression flange is subject to torsional buckling over the support owing to its unrestrained condition, and thus the buckling resistance moment is reduced. The length of the unrestrained portion may be considered as one-quarter of the span on either side of the support, i.e. the point of contraflexure in the bending moment diagram. So, let us consider the case when wind pressure acts on the member along its major axis. The buckling resistance moment may be calculated in the following way. The length between the restraints of the compression flange over the support is

$$L_e = 1/4 \times 6000 = 1500\ \text{mm}$$

Referring to the equation in Clause 6.3.2.4,

$$\bar{\lambda}_f = k_c L_c/(i_{f,z}\lambda_1) \leq \lambda_{c0}(M_{c,Rd}/M_{y,Ed}) \quad (6.59)$$

As already calculated in the design of the purlin, $M_{c,Rd} = 171\ \text{kN m} > M_{Ed}$ (18.9 kN m). So, OK.

Along the *minor axis* of the member, when a vertical dead load is acting,

$$M_{uz} = 3.1\ \text{kN m}$$

Buckling moment of resistance $= M_{b,Rd}(z) = W_z f_y/\gamma_{M1} = 142 \times 10^3 \times 275/10^6 = 39.1\ \text{kN m} > M_{uz}$ (3.1 kN m)

Therefore

$$M_{uy}/M_{b,Rd}(y) + M_{uz}/M_{b,Rd}(z) = 18.9/171 + 3.1/39.1 = 0.11 + 0.08 = 0.19 < 1$$ OK

Therefore, we adopt UB305 × 165 × 40 kg/m for the side rail.

5.3 Design of roof trusses (members subjected to compression and tension)

5.3.1 Design considerations

The roof trusses are to be designed as simply supported at their ends on the roof columns and are subjected to dead, live and wind loads.

As no snow loads are considered, the roof slope has been assumed to be low. This will allow the eaves to end at a depth of about 2.5 m, and will form the roof truss into a stiff frame that will transfer wind forces to the adjacent frame and at the same time will be stable enough resist the high transverse forces generated during the movement of very heavy overhead electric travelling cranes. The slope of the roof should be kept to more than 6° (the minimum permissible slope).

The roof truss forms an *N*-member system as shown in Fig. 5.3. The members may be subjected to reversal of stresses owing to various loading conditions. So, the members should be designed for compression forces as well as for tension forces.

5.3.2 Design data

Span = 27.0 m.

Spacing = 6.0 m (maximum).

Spacing of node points = 2.25 m.

5.3.3 Loadings, based on Eurocode 1, Part 1-1

5.3.3.1 Dead loads

We assume 20G corrugated galvanized steel sheeting, including side laps two corrugations wide, and 150 mm end laps.

Load due to sheeting = 0.11 kN/m^2.

Insulation (10 mm fibre board) = 28 N/m^2 = 0.028 kN/m^2.

Service loads (including lighting, sprinklers etc.) = 0.1 kN/m^2.

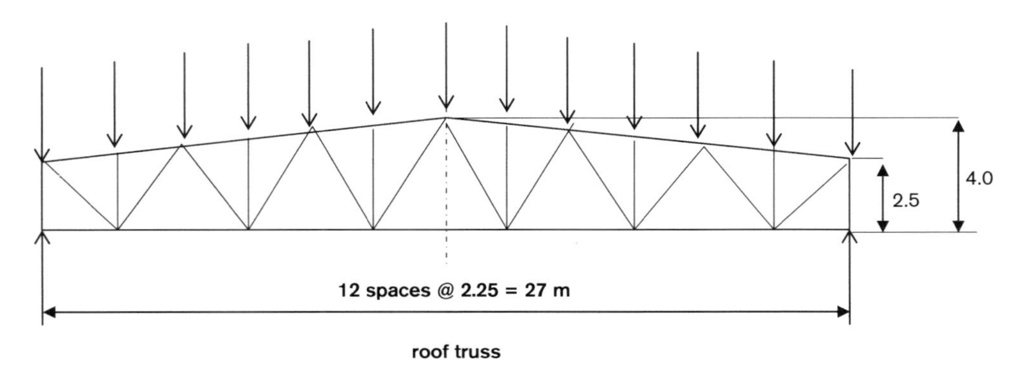

Fig. 5.3. Arrangement of members

Weight of purlin (UB203 × 133 × 25 kg/m) = 25 × 6/(2.25 × 6) = 11 kg/m^2 = 0.11 kN/m^2.

Self-weight of truss (assumed) = 0.20 kN/m^2.

Total = 0.548 kN/m^2.

Therefore

load/m on trusses with 6 m spacing = g_k = 0.548 × 6 = 3.288 kN/m;

load per node point (spacing = 2.25 m) = G_k = 3.288 × 2.25 = 7.398, say 7.4 kN.

5.3.3.2 Live loads

For a roof slope not greater than 10°, with access, the imposed load q_k is 1.5 kN/m^2.

Load/m run of truss with 6.0 m spacing = 1.5 × 6 = 9 kN/m.

Therefore

imposed load per node point = Q_k = 9 × 2.25 = 20.25 kN.

5.3.3.3 Wind loads, based on Eurocode 1, Part 1-4

Wind velocity pressure (dynamic) = $q_p(z)$ = 1.27 kN/m^2

(previously calculated in Chapter 2). When the wind is blowing from right to left, the resultant pressure coefficient on the windward and leeward slopes with positive internal pressure is

$c_{pe} = -0.9$ (already calculated)

Therefore the external wind pressure normal to the roof is

$q_e = q_p c_{pe} = -1.27 \times 0.9 = 1.14$ kN/m^2

Vertical component $p_{ev} = p_e \cos\theta = 1.43 \times \cos 6.28 = 1.13$ kN/m^2 acting upwards ↑

Therefore the total vertical component of the wind force at a node point is

$W_k = 1.13 \times 2.25 \times 6 = 15.3$ kN $> G_k$.

So, there is the possibility of reversal of stresses in the members.

Now, we have to consider several load combinations as follows.

5.3.4 Forces in members

Before we analyse the forces, the following points should be noted:

- For equilibrium, the sum of all vertical forces at a node must be zero, i.e. $\sum V = 0$, and the sum of all horizontal forces at the node must be zero, i.e. $\sum H = 0$.
- The type of force in a member will be indicated in illustrations by arrows at each end. Arrows directed outwards denote compression in the member, and arrows directed inwards denote tension.

- In the analysis of the action of forces in members meeting at a node point, the forces in the members will be resolved into horizontal and vertical components. If an arrow symbolizing the force in a member is directed outwards from a node, this means that the member is in tension, and the arrow is directed inwards towards the node, this indicates that the member is in compression.

We follow Fig. 5.4 (note that the forces in the members are unfactored), for dead loads only. *Consider node 1.*

Length of member (1-9) = $(2.5^2 + 2.25^2)^{0.5} = 3.36$ m.

Let θ be the inclination of the top chord to the horizontal:

$\tan\theta = 1.5/13.5 = 0.11;\ \theta = 6.28°$.

Take moments about node 9:

(1-2) = $[(44.4 \times 2.25 - 3.7 \times 2.25)/2.75] \times (2.264/2.25) = 33.3 \times 2.264/2.26 = 33.5$ kN (compression).

$\sum V = 0: \quad -3.7 + 44.4 - (1\text{-}9) \times 2.5/3.36 - (1\text{-}2) \times 0.25/2.264 = 0.$

Therefore

(1-9) = 59.6 kN (tension).

Consider node 9.

$\sum V = 0: \quad 59.6 \times 2.5/3.36 - 7.4 - (3\text{-}9) \times 3/3.75 = 0.$

Therefore

(3-9) = $36.945 \times 3.75/3 = 46.2$ kN (compression).

$\sum H = 0: \quad -59.6 \times 2.25/3.36 - 7.4 - (3\text{-}9) \times 2.25/3.75 + (9\text{-}10) = 0.$

Therefore

(9-10) = 57.6 (tension).

Consider node 10.

$(3\text{-}4) \times 3.25 = [44.4 \times 6.75 - 3.7 \times 6.75 - 7.4 \times (2.25 + 4.5)] \times (2.264/2.25).$

Therefore

(3-4) = 69.6 kN (compression);

(4-5) = 69.6 kN (compression).

Consider node 5.

$(10\text{-}11) \times 3.5 = 44.4 \times 9 - 3.7 \times 9 - 7.4 \times (2.25 + 4.5 + 6.75) = 266.4.$

Therefore

(10-11) = $266.4/3.5 = 76.1$ kN (tension).

Consider node 3.

$\sum V = 0$:
$-7.4 + 46.2 \times 3/3.75 - (3\text{-}10) \times 3/3.75 + 33.5 \times 0.25/2.264 - 69.6 \times 0.25/2.264 = 0.$

Therefore

$(3\text{-}10) = 25.57 \times 3.75/3 = 32$ kN (tension).

Consider node 10.

$(3\text{-}10) \times 3/3.75 - 7.4 - (5\text{-}10) \times 3.5/4.16) = 0.$

Therefore

$(5\text{-}10) = 21.6$ kN (compression).

Consider node 11.

$(5\text{-}6) \times 3.75 = 44.4 \times 11.25 - 3.7 \times 11.25 - 7.4(2.25 + 4.5 + 6.75 + 9).$

Therefore

$(5\text{-}6) = 78.2$ kN (compression).

Consider node 5.

$\sum V = 0$:
$(5\text{-}10) \times 3.5/4.16 - (5\text{-}11) \times 3.5/4.16 + (4\text{-}5) \times 0.25/2.264 - (5\text{-}6) \times 0.25/2.264 - 7.4 = 0.$

Therefore

$(5\text{-}11) = 9.82 \times 4.16/3.5 - 7.4 = 11.67$ kN (tension).

Consider node 7.
Take moments about node 7:

$(11\text{-}12) \times 4 = 44.4 \times 13.5 - 3.7 \times 13.5 - 7.4(2.25 + 4.5 + 6.75 + 9 + 11.25).$

Therefore

$(11\text{-}12) = 299.7/4 = 74.9$ kN (tension).

Consider node 11.

$\sum V = 0$: $(5\text{-}11) \times 3.5/4.16 - 7.4 + (7\text{-}11) \times 4/4.59 = 0.$

Therefore

$(7\text{-}11) = 2.42 \times 4.59/4 = 2.8$ kN (tension).

$\sum H = 0$: $-76.1 - (5\text{-}11) \times 2.25/4.16 + (7\text{-}11) \times 2.25/4.59 + (11\text{-}12) = 0.$

Therefore

(11-12) = 81 kN (tension).

(See Fig. 5.4, showing the loadings and characteristic (unfactored) forces in the members.)

5.3.5 Load combinations for ultimate design force in the members by ULS method

Referring to Table A1.2(B) of EN 1990: 2002(E), we consider three cases as follows.

Case 1: when only DL + LL are acting, no wind load.

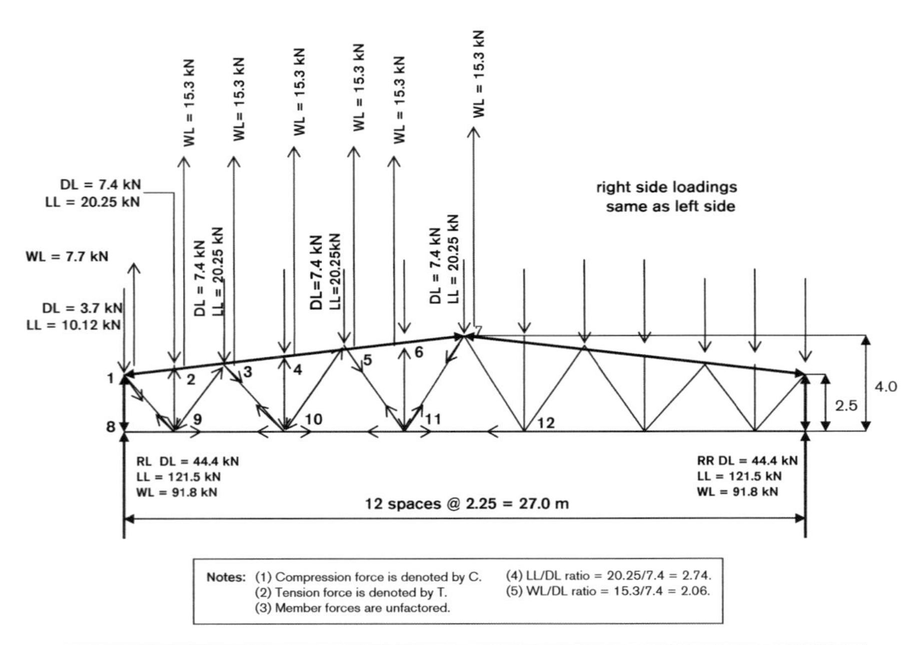

Member	Location	Force DL (kN)	Force LL (kN)	Force WL (kN)	Member	Location	Force DL (kN)	Force LL (kN)	Force WL (kN)
(1–2)	top chord	33.5C	91.8C	69.0T	(1–9)	diagonal	59.6T	163.3T	122.8C
(2–3)	top chord	33.6C	91.8C	69.0T	(3–9)	diagonal	46.2C	126.6C	95.2T
(3–4)	top chord	69.6C	190.7C	143.4T	(3–10)	diagonal	32.0T	87.7T	65.9C
(4–5)	top chord	69.6C	190.7C	143.4T	(5–10)	diagonal	21.6C	59.2C	44.5T
(5–6)	top chord	78.2C	214.3C	161.1T	(5–11)	diagonal	11.7T	32.0T	24.1C
(6–7)	top chord	78.2C	214.3C	161.1T	(7–11)	diagonal	2.8T	7.7T	5.8C
(8–9)	bottom chord	0.0	0.0	0.0	(1–8)	vertical	44.4C	121.7C	91.5T
(9–10)	bottom chord	57.6T	157.8T	118.7C	(2–9)	vertical	7.4C	20.3C	15.2T
(10–11)	bottom chord	76.1T	208.5T	156.8C	(4–10)	vertical	7.4C	20.3C	15.2T
(11–12)	bottom chord	81.0T	221.9T	166.9C	(6–11)	vertical	7.4C	20.3C	15.2T

Fig. 5.4. Roof truss loadings and forces in the members, unfactored

Partial factor for permanent actions (DL) = γ_{Gj} = 1.35 (unfavourable).

Partial factor for leading variable actions (LL) = γ_{Qk} = 1.35.

Therefore

ultimate design force = $F_u = \gamma_{Gj} G_k + \gamma_{Qk} Q_k = 1.35 G_k + 1.5 Q_k$.

Case 2: when DL + LL + WL are acting simultaneously. We use a partial factor for the accompanying variable actions of wind loads equal to $\gamma_{Wk}\psi_0 = 1.5 \times 0.6 = 0.9$ (for the value of ψ_0, refer to Table A1.1 of BS EN 1990: 2002(E) (Eurocode, 2002b). Therefore the ultimate design force in the member is

$$F_u = \gamma_{Gj} G_k + \gamma_{Qk} Q_k + \gamma_{Wk} \psi_0 W_k = 1.35 G_k + 1.5 Q_k + 0.9 W_k.$$

Case 3: when only DL + WL are acting.

Partial factor for permanent actions (DL) = γ_{Gj} = 1.0 (favourable).

Partial factor for leading variable actions (WL) = γ_{Wk} = 1.5.

Therefore

ultimate design force in the member = $F_u = \gamma_{Gj} G_k + \gamma_{Wk} W_k = G_k + 1.5 W_k$.

Thus, we conclude that in order to get the maximum ultimate design force in the member, we have to check the above three cases.

5.3.6 Design of section of members, based on Eurocode 3, Part 1-1

(See Fig. 5.4 for the unfactored forces in the members.) We consider the chord members first.

5.3.6.1 Top chords

Design data

Consider the member (6-7). The unfactored forces are

DL = 78.2 kN, LL = 214.3 kN, WL = 161.3 kN.

The ultimate design force in the member is calculated as follows.

Case 1: when DL + LL are acting, no wind load.

Ultimate design force = N_{Ed} = (1.35 × 78.2 + 1.5 × 214.3) = 427 kN (compression).

Case 2: when DL + LL + WL are acting simultaneously.

Ultimate design force = N_{Ed} = (1.35 × 78.2 + 1.5 × 214.3 − 0.9 × 161.3 = 281 kN (compression).

Case 3: when only DL + WL are acting.

Ultimate design force = N_{Ed} = 1.0 × 78.2 − 1.5 × 161.1 = −163.5 (tension).

Thus we find that the member is subject to reversal of stresses. So, we have to design the member as a compression member and check it also as a tension member.

To design the member as a compression member

Try a section made from two angles 100 × 100 × 8 back to back with a 10 mm gap between the vertical faces.

$A = 31$ cm², $iy = 3.06$ cm, $iz = 4.46$ cm.

The classification of the cross-section is done as follows:

$f_y = 275$ N/mm²; $e = (235/f_y)^{0.5} = 0.92$;

$h/t = 100/8 = 12.5$; $15\varepsilon = 13.8$.

Referring to Table 5.2 (sheet 3) of Eurocode 3, Part 1-1, for class 3 classification,

$$h/t \leq 15\varepsilon \quad \text{and} \quad (h + b)/2t \leq 11.5\varepsilon$$

We have 12.5 < 13.8. So, the cross-section satisfies the first condition. We also have

$$(h + b)/2t = (100 + 100)/2 \times 8 = 12.5$$

where b is the depth of the vertical leg, and

$$11.5\varepsilon = 11.5 \times 0.92 = 10.6$$

but in our case 12.5 > 10.6. This does not satisfy the conditions. So, we increase the thickness of the angle from 8 mm to 10 mm, and try two angles 100 × 100 × 10 back to back with a 10 mm gap (see Fig. 5.5). We have

$A = 38.4$ cm², $iy = 3.04$ cm, $(h + b)/2t = 10$, $11.5\varepsilon = 10.6$.

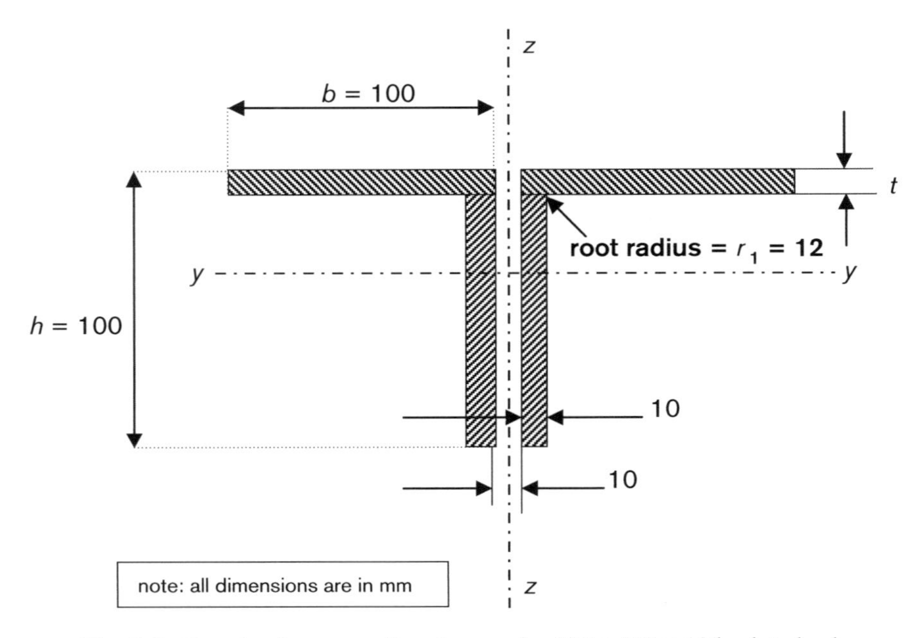

Fig. 5.5. Top chord cross-section, two angles 100 × 100 × 10 back to back

For class 3 classification, $(h + b)/2t \leq 11.5\varepsilon$. We have $10.0 < 10.6$. This satisfies the conditions.
OK

The buckling resistance to compression is checked as follows. Referring to Clause 6.3.1, a compression member should be verified against buckling by the following equation:

$$N_{Ed}/N_{b,Rd} \leq 1.0 \quad (6.46)$$

where N_{Ed} is the ultimate design compressive force and $N_{b,Rd}$ is the design buckling resistance of the compression member. The design buckling resistance of a compression member is given by the following equation:

$$N_{b,Rd} = \chi A f_y/\gamma_{M1} \quad (6.47)$$

where χ is the reduction factor for the relevant buckling mode, A is the gross area of the section = 38.4 cm^2, f_y = 275 N/mm^2, γ_{M1} is the partial factor, equal to 1 (see Clause 6.1, note 2B), and

$$\chi = 1/(\Phi + (\Phi^2 - \bar{\lambda}^2)^{0.5}) \quad (6.49)$$

where $\Phi = 0.5[1 + \alpha(\bar{\lambda} - 0.2) + \bar{\lambda}^2]$

We now use the equation

$$\bar{\lambda} = L_{cr}/(iy\lambda_1), \quad (6.50)$$

where

$\lambda_1 = 93.9\varepsilon = 93.9 \times 0.92 = 86.4$,
L_{cr} = buckling length in the buckling plane = 2250 mm,
i_y = radius of gyration about y–y axis = 3.04 cm.

Therefore

$$\bar{\lambda} = 2250/(30.4 \times 86.4) = 0.86$$

Referring to Table 6.2 ("Selection of buckling curve for a cross-section") of Eurocode 3, Part 1-1 (see Appendix B), in our case of an L-section, we follow the buckling curve "b" in Fig. 6.4 of that Eurocode. With $\bar{\lambda} = 0.86$, $\chi = 0.7$. Also, applying equation (6.49),

$$\Phi = 0.5[1 + \alpha(\bar{\lambda} - 0.2) + \bar{\lambda}^2$$

where α is an imperfection factor. Referring to Table 6.1 ("Imperfection factors for buckling curves") of Eurocode 3, Part 1-1, with the buckling curve "b" in Fig. 6.4 of that Eurocode, $\alpha = 0.34$. We obtain

$$\Phi = 0.5[1 + 0.34(0.86 - 0.2) + 0.86^2] = 0.5 \times 1.96 = 0.98$$
$$\chi = 1/[0.98 + (0.98^2 - 0.86^2)^{0.5}] = 1/1.45 = 0.72.$$

Therefore

$$N_{b,Ed} = \chi A f_y/\gamma_{M1} = 0.7 \times 38.4 \times 100 \times 275/103 = 739 \text{ kN};$$

$N_{Ed}/N_{b,Rd} = 427/739 = 0.58 < 1.$ Satisfactory

The ultimate tension force is much less than the compressive force. It is found to be quite satisfactory. Therefore, we adopt two angles 100 × 100 × 10 back to back with a 10 mm vertical gap for the top chord.

5.3.6.2 Bottom chords

We consider the member (11-12).

Design data

DL = 81 kN (tension),

LL = 221.9 kN (tension),

WL = 166.9 kN (compression).

The ultimate design force in the member is calculated as follows.

Case 1: when DL + LL are acting, no WL.

Ultimate design force = (1.35 × 81 + 1.5 × 221.9) = 442.2 kN (tension).

Case 2: when DL + LL + WL are acting simultaneously.

Ultimate design force = (1.35 × 81 + 1.5 × 221.9 + 0.9 × 166.9) kN = 292 kN (tension).

Case 3: when DL + WL are acting, no LL.

Ultimate design force = (1.0 × 81 − 1.5 × 166.9) kN = −169.4 kN (compression).

Thus the member is subject to reversal of stresses. So, we have to design the member for tension and to check if the member is capable of resisting the compression force.

Design of section

Maximum ultimate tension = 442.2 kN.

Try two angles 2 × 100 × 100 × 8 back to back with a 10 mm gap between the vertical faces.

$A_{gross} = 31.0\ \text{cm}^2$

Using 20 mm diameter bolts, $A_{net} = A_{gross} - 2 \times 2.2 \times 0.8 = 27.5\ \text{cm}^2$. The design plastic resistance is given by

$$N_{pl,Ed} = A_{gross} f_y/\gamma_{Mo} \qquad (6.6)$$

$$= 31 \times 102 \times 275/103 = 852.5\ \text{kN}$$

and the design ultimate resistance is given by

$$N_{u,Rd} = 0.9 A_{net} f_u \qquad (6.7)$$

$$= 0.9 \times 27.7 \times 430/10 = 1072\ \text{kN}$$

The smaller of the above should be taken as the design tension resistance = 852.5 > N_{Ed} (the design tension). Satisfactory

Therefore, we adopt two angles 100 × 100 × 10 back to back with a 10 mm gap between the vertical faces (since a member made from two angles 100 × 100 × 8 does not satisfy the section classification).

Now, we must verify if the member is capable of resisting a compression equal to 169.4 kN. The design buckling resistance is $\chi A f_y/\gamma_{M1}$, $\bar{\lambda} = L_{cr}/iy_1$, $\lambda_1 = 86.4$, L_{cry} = buckling length = 4500 mm and iy = 3.06 cm. Hence

$$\bar{\lambda} = 4500/(30.6 \times 86.4) = 1.7$$

Referring to Table 6.2 of Eurocode 3, Part 1-1, for our L-section, we follow the buckling curve "b" in Fig. 6.4 of that Eurocode. With $\bar{\lambda} = 1.7$, $\chi = 0.28$. Therefore

$N_{b,Rd} = \chi A f_y/\gamma_{M1} = 0.28 \times 31 \times 275/10 = 238.7$ kN (compression) > N_{Ed} (169.4 kN) (compression) Satisfactory

5.3.6.3 Diagonals and verticals

Diagonal member (3-9)

The design data are as follows. The unfactored forces are

DL = 42.6 kN (compression),

LL = 126.6 kN (compression),

WL = 95.2 kN (tension).

For the ultimate design forces in the member, we consider three cases.

Case 1: when DL + LL are acting.

Ultimate design force = (1.35 × 46.2 + 1.5 × 126.6) = 247.4 kN (compression).

Case 2: when DL + LL + WL are acting simultaneously.

Ultimate design force = (1.35 × 46.2 + 1.5 × 126.6 + 0.9 × 95.2) kN = 338 kN (compression).

Case 3: when DL + WL are acting, no LL.

Ultimate design force = (1.35 × 46.2 − 1.5 × 95.2) kN = −80.43 kN (tension).

Thus, the diagonals are subjected to reversal of stresses. We design the member as a compression member, with N_{Ed} = 247.4 kN (compression). For the design of the section, we try two angles 100 × 100 × 10 back to back:

$$(h + b)/2t = 180/16 = 11.25;\ 11\varepsilon = 10.6$$

Table 5.2 (sheet 3) of Eurocode 3, Part 1-1, specifies the condition $(h + b)/2t \leq 11\varepsilon$. In our case, the section does not satisfy this condition.

Therefore, we adopt for all diagonals a higher thickness of two angles 90 × 90 × 10 back to back with a 10 mm gap between the vertical faces.

Verticals

We consider first the vertical member (6-11). The design data are as follows. The unfactored forces are

DL = 7. 4 kN (compression),

LL = 20.3 kN (compression),

WL = 15.2 kN (tension).

Case 1: when DL + LL are acting.

Ultimate design force = (1.35 × 7.4 + 1.5 × 20.3) kN = 40.4 kN (compression).

Case 3: when DL + WL are acting, no LL.

Ultimate design force = (7.4 − 1.5 × 15.2) kN = −15.4 (tension).

We design the member as a compression member, with N_{Ed} = 40.4 kN. For the design of the section, we try two angles 70 × 70 × 8 back to back with a 10 mm gap between the vertical faces:

$(h + b)/2t = (70 + 70)/(2 \times 8) = 8.75,$

$11\varepsilon = 10.12$

Table 5.2 (sheet 3) of Eurocode 3 specifies the condition $(h + b)/2t < 11\varepsilon$. So, the section satisfies this condition.

Therefore we adopt two angles 70 × 70 × 8 back to back with a 10 mm gap between the vertical faces for all vertical members except member (1-8).

Next, we consider the member (1-8). The design data are as follows. The unfactored forces are

DL = 44.4 kN (compression),

LL = 121.7 kN (compression),

WL = 91.5 kN (tension).

For the ultimate design forces in the member, we consider two cases.

Case 1: when DL + LL are acting, no wind load.

Ultimate design force = (1.35 × 44.4 + 1.5 × 121.7) kN = 242.5 kN (compression).

Case 2: when DL = WL are acting no LL.

Ultimate design force = (44.4 − 1.5 × 121.7) kN = −138.2 kN (tension).

Thus, the vertical members are also subject to reversal of stresses. We design the member as a compression member, with N_{Ed} = 242.5 kN (compression). For the design of the section, we try two angles 100 × 100 × 10 back to back:

$(h + b)/2t = 10$ and $11\varepsilon = 10.6$.

Table 5.2 (sheet 3) of Eurocode 3, Part 1-1, specifies the condition $(h + b)/2t < 11\varepsilon$. So, the section satisfies this condition.

Therefore we adopt two angles 100 × 100 × 10 back to back with a 10 mm gap between the vertical faces (see Fig. 5.6, showing the ultimate design forces in the members and the sizes of the members).

5.4 Roof girders in melting bay (members subjected to compression and tension)

5.4.1 Design considerations

The main stanchions in the melting bay are spaced at 24.0 m (maximum). The spacing of roof trusses has been selected as 6.0 m (maximum). So, the intermediate roof trusses between the main stanchions should be supported by roof girders at roof level, with spans

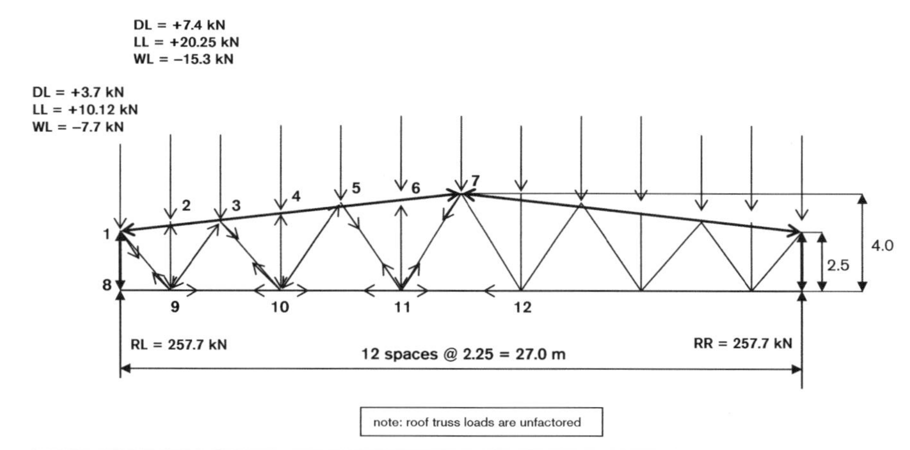

Member	Location	Forces (kN) (ultimate)	Member size 2 b to b angles	Member	Location	Forces (kN) (ultimate)	Member size 2 b to b angles
(1–2)	top chord		2 × 100 × 100 × 10	(1–9)	diagonal		2 × 90 × 90 × 10
(2–3)	top chord			(3–9)	diagonal	247.4 comp. 80.4 tension	2 × 90 × 90 × 10
(3–4)	top chord			(3–10)	diagonal		2 × 90 × 90 × 10
(4–5)	top chord			(5–10)	diagonal		2 × 90 × 90 × 10
(5–6)	top chord			(5–11)	diagonal		2 × 90 × 90 × 10
(6–7)	top chord	427.0 comp. 163.5 tension		(7–11)	diagonal		2 × 90 × 90 × 10
(8–9)	bottom chord	0.0		(1–8)	vertical	24.5 comp. 138.2 tension.	2 × 100 × 100 × 10
(9–10)	bottom chord			(2–9)	vertical		2 × 70 × 70 × 8
(10–11)	bottom chord			(4–10)	vertical		2 × 70 × 70 × 8
(11–12)	bottom chord	442.2 tension 169.4 comp.		(6–11)	vertical	40.4 comp. 15.4 tension	2 × 70 × 70 × 8

Fig. 5.6. Melting shop, showing ultimate design forces in roof truss members and sizes of members

of 24.0 m between the main stanchions. The roof girders may be of plate girder or lattice-type girder construction. The supports are considered to be simply supported. The depth of a girder may be assumed to be 1/10 to 1/12 of the span. See Figs 5.7 (elevation) and 5.8 (plan), showing the arrangement of the roof girders, intermediate columns and horizontal roof bracing system.

5.4.2 Functions

In our case, the roof girders support three intermediate roof trusses. The roof girders transmit all of the vertical loads from the intermediate roof trusses to the main stanchions. The roof girders also support intermediate columns at roof level, which are spaced at 6.0 m between the main stanchions. The horizontal reaction of the columns at the top due to wind is transferred to the roof girders. To resist these horizontal loads, a horizontal braced

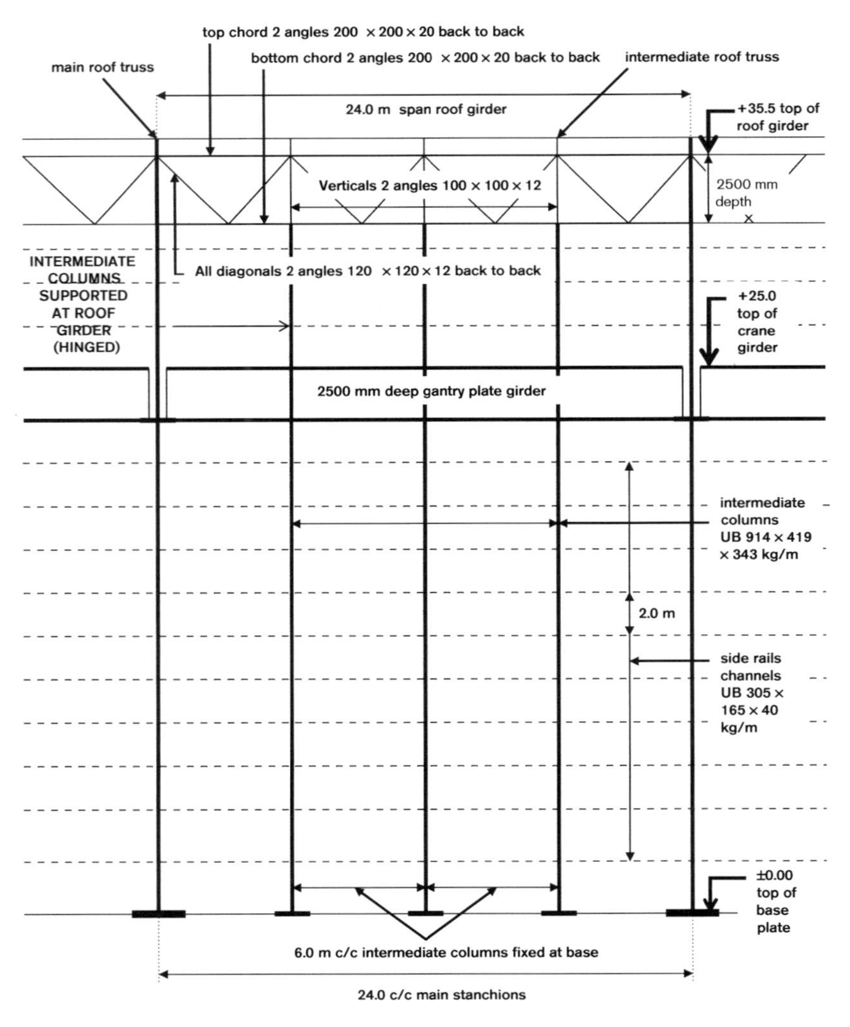

Fig. 5.7. Elevation, showing arrangement of roof girders and intermediate columns

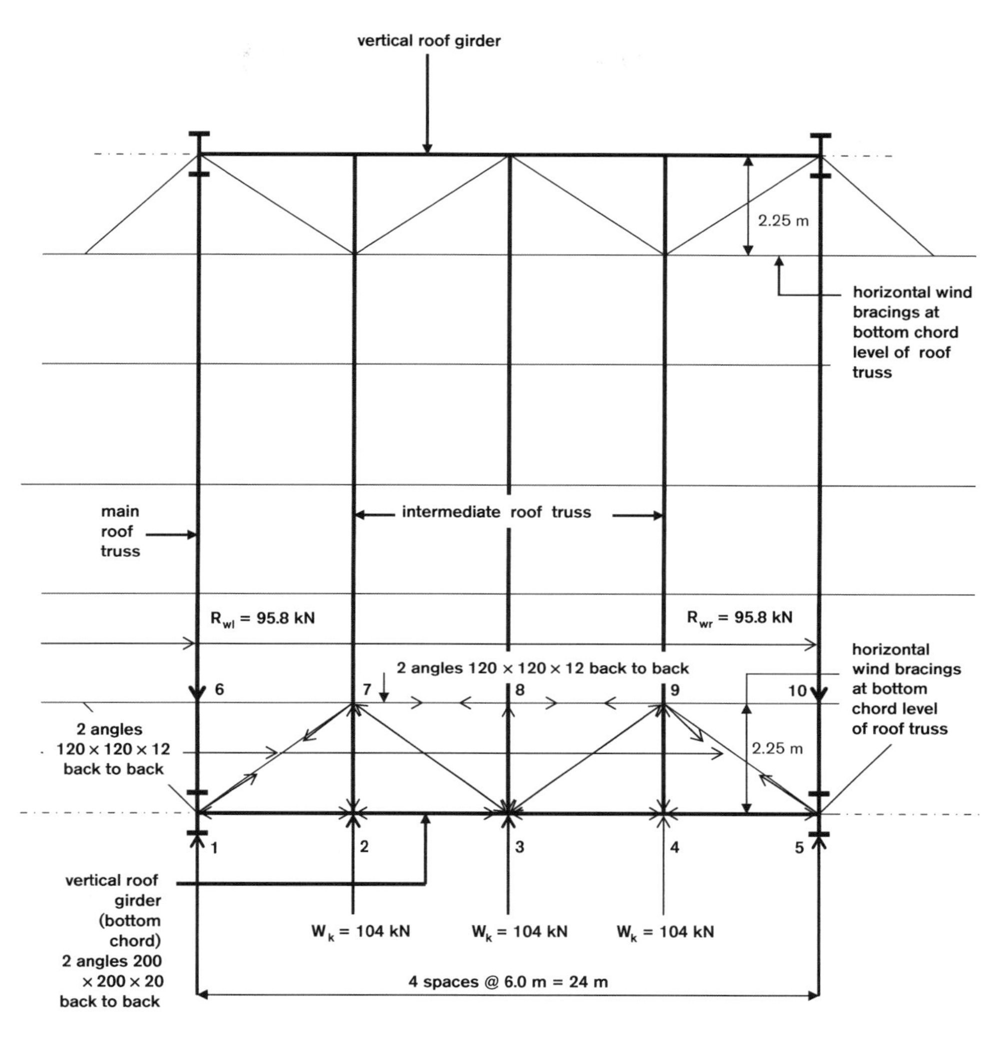

Fig. 5.8. Plan at bottom chord level of roof truss, showing horizontal roof bracing system

girder of 24 m span, connected to the roof girders spanning between the main stanchions, was considered. See Figs 5.7 and 5.8, which show a vertical roof girder and a horizontal braced wind girder.

5.4.3 Design data

Span = 24.0 m.

Vertical depth (assumed) = 1/10 of span = 24/10 = 2.40 m.

The girder is subjected to vertical and horizontal loads.

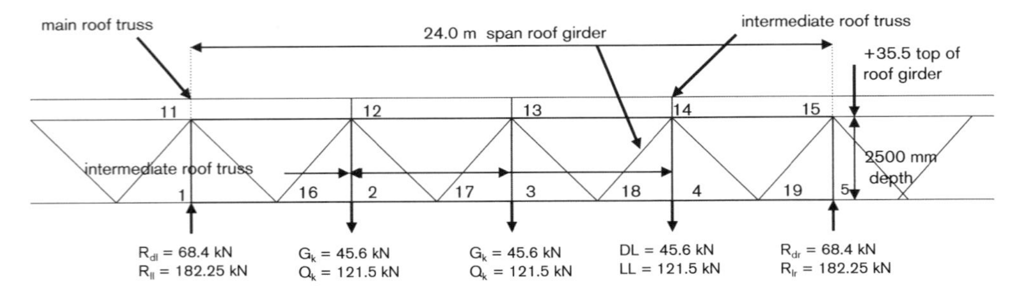

Fig. 5.9. Loadings on latticed roof girder

5.4.4 Loadings, based on Eurocode 1, Part 1-1

5.4.4.1 Vertical dead loads

Reaction from intermediate truss $R_d = G_k = 3.288 \times 27/2 = 44.4$ kN.

Self-weight/m = 0.2 kN/m × 6 per node (6 m spacing) = 1.2 kN.

Total = 45.6 kN.

5.4.4.2 Vertical live loads

Reaction from intermediate truss $R_l = Q_k = 9 \times 27/2 = 121.5$ kN
(see Fig. 5.9, showing loadings on the latticed roof girder.)

5.4.4.3 Horizontal wind loads: reaction from intermediate column

From previous calculations, the effective wind pressure normal to the wall face is

$$p_e = q_p(z) \times \text{resultant pressure coefficient} = q_p(z)c_{pe} = 1.27 \times 1.1 = 1.4 \text{ kN/m}^2$$

where $q_p(z)$ is the dynamic wind pressure = 1.27 kN/m^2 and c_{pe} is the resultant pressure coefficient on the windward vertical face = 1.1. (All of these values have been calculated earlier.)

Spacing of intermediate columns = 6.0 m.

Wind force/m height of column = $w_k = 1.4 \times 6 = 8.4$ kN/m.

Assume that the intermediate column is fixed at the base and freely supported at the top. Therefore the horizontal reaction at the top/node is

$R_w = (3/8)w_k h$ (where h = height of column = 33 m (underside of truss))

$= 3/8 \times 8.4 \times 33 = 104$ kN.

(The reactions calculated here are unfactored.)

Unfactored vertical reaction due to dead load/node = $R_d = 45.6$ kN.

Unfactored vertical reaction due to imposed load/node = $R_l = 121.5$ kN.

Unfactored horizontal reaction due to wind/node = $R_w = 104$ kN.

5.4.5 Forces in members due to unfactored dead loads

See Fig. 5.9, showing the loadings.

Reaction $R_1 = 45.6 \times 1.5 = 68.4$ kN.

Consider node 11.

$\sum V = 0$: length of member (2-5) $= (2.5^2 + 3.0^2)^{0.5} = 3.9$ m;

$(11\text{-}16) \times 2.5/3.9 = 68.4$.

Therefore

$(11\text{-}16) = 68.4 \times 3.9/2.5 = 106.7$ kN (tension).

$\sum H = 0$.

Therefore

$(11\text{-}12) = 106.7 \times 3.9/3.0 = 106.7 \times 3/3.9 = 82.1$ kN (compression).

Next, consider node 16.

$\sum V = 0$.

Therefore

$(12\text{-}16) = (11\text{-}16) = 106.7$ kN (compression).

$\sum H = 0$.

Therefore

$(2\text{-}16) = (11\text{-}16) \times 3/3.9 - (12\text{-}16) \times 3/3.9 = 2 \times 106.7 \times 3/3.9 = 164.2$ kN (tension).

Next, consider node 12.

$\sum V = 0$.

Therefore

$+(12\text{-}16) \times 2.5/3.9 - (12\text{-}17) \times 2.5/3.9 - 45.6 = 0$;

$(12\text{-}17) = 35.6$ kN (tension).

$\sum H = 0$.

Take moments about node 17. Therefore

$(12\text{-}13) = [68.4 \times 9 - 45.6 \times 3]/2.5 = 191.5$ kN (compression).

Next, consider node 17.

$\sum H = 0$:

$(3\text{-}17) = (12\text{-}17) \times 3/3.9 + (13\text{-}17) \times 3/3.9 + 164.2 = 219$ kN (tension).

5.4.6 Forces due to unfactored imposed loads

The multiplying ratio LL/DL is equal to 121.5/45.6 = 2.664. Therefore the forces in the members are as follows:

top chord, (11-12) = 82.1 × 2.664 = 218.7 kN (compression);

top chord, (12-13) = 191.6 × 2.664 = 510.4 kN (compression);

bottom chord, (1-5) = 0;

bottom chord, (2-16) = 164.2 × 2.664 = 437.4 kN (tension);

bottom chord, (2-17) = 164.2 × 2.664 = 437.4 kN (tension);

bottom chord, (3-17) = 219 × 2.664 = 583.4 kN (tension);

diagonal, (11-16) = 106.7 × 2.664 = 284.3 kN (tension);

diagonal, (12-16) = 106.7 × 2.664 = 284.3 kN (compression);

diagonal, (12-17) = 35.6 × 2.664 = 94.8 kN (tension);

diagonal, (13-17) = 35.6 × 2.664 = 94.8 kN (compression);

vertical, (1-11) = 68.4 × 2.664 = 182.2 kN (compression);

vertical, (2-12) = 45.6 × 2.664 = 121.5 kN (tension).

5.4.7 Ultimate forces in members due to (DL + LL) without wind

Referring to BS EN 1990: 2002(E), and using partial safety factors $\gamma_f = 1.35$ and 1.5 for the load combinations DL and LL, respectively, the ultimate forces in the members due to DL + LL are as given below, in the form "ultimate DL + ultimate LL = total".

Top chord:

(11-12) = 1.35 × 82.1 + 1.5 × 218.7 = 439 kN (compression);

(12-13) = 1.35 × 191.6 + 1.5 × 510.4 = 1024 kN (compression).

Bottom chord:

(1-5) = 0.0 + 0.0 = 0.0;

(2-16) = 1.35 × 164.2 + 1.5 × 437.4 = 878 kN (tension);

(2-17) = 1.35 × 164.2 + 1.5 × 437.4 = 878 kN (tension);

(3-17) = 1.35 × 219.0 + 1.5 × 583.4 = 1171 kN (tension).

Diagonals:

(11-16) = 1.35 × 106.7 + 1.5 × 284.3 = 570 kN (tension);

(12-16) = 1.35 × 106.7 + 1.5 × 284.3 = 570 kN (compression);

(12-17) = 1.35 × 35.6 + 1.5 × 94.8 = 190 kN (tension);

(13-17) = 1.35 × 35.6 + 1.5 × 94.8 = 190 kN (compression).

5.4.8 Design of section of members, based on Eurocode 3, Part 1-1

5.4.8.1 Top chord (12-13)

Ultimate design force = 1024.0 kN (compression).

Initial sizing of member

Try two angles 150 × 150 × 18 back to back: $A_g = 102\ \text{cm}^2$, $i_y = 4.56$ cm, $l = 600$ cm.

Classification of section

$f_y = 275\ \text{N/mm}^2$, $\varepsilon = 0.92$,

$h/t = 150/18 = 8.3$.

Referring to Table 5.2 (sheet 3) of Eurocode 3, Part 1-1, for class 3 classification, $h/t \leq 15\varepsilon$ and $(h + b)/2t \leq 11.5\varepsilon$. In our case,

$15\varepsilon = 15 \times 0.92 = 13.8 > h/t\ (8.3)$ <u>OK</u>

$(h + b)/2t = 8.3 < 10.8\ (11.5 \times 0.92)$ <u>OK</u>

Thus, the section satisfies both of the conditions.

Buckling resistance to compression

Referring to Clause 6.3.1 of Eurocode 3, Part 1-1, a compression member should be verified against buckling by the following equation:

$$N_{Ed}/N_{b,Rd} \leq 1.0 \qquad (6.46)$$

The design buckling resistance of a compression member is given by the following equation:

$$N_{b,Rd} = \chi A f_y/\gamma_{M1} \qquad (6.47)$$

where

$\bar{\lambda}$ = non-dimensional slenderness = $L_{cr}/(i_y\lambda_1)$,
L_{cr} = buckling length in the buckling plane considered = 6,
$\lambda_1 = 93.9\varepsilon = 93.9 \times 0.92 = 86.4$,
$\bar{\lambda} = 600/(4.56 \times 86.4) = 1.52$.

Referring to Table 6.2 ("Selection of buckling curve for a cross-section") of Eurocode 3, Part 1-1, for any axis of an angle section, we follow the buckling curve "b" in Fig. 6.4 of that Eurocode. With $\bar{\lambda} = 1.51$, the reduction factor $\chi = 0.34$. Therefore

$N_{b,Rd} = 0.34 \times 102 \times 100 \times 275/103 = 954\ \text{kN} < N_{Ed}$ (1024 kN), not acceptable.

So, we increase the section to two angles 200 × 200 × 20 back to back (see Fig. 5.10). We have

$A = 153\ \text{cm}^2$, $i_y = 6.11$, $L_{cr} = 600$ cm,

$(h + t)/2t = 10$, $11.5\varepsilon = 11.5 \times 0.92 = 10.6$.

So, $(h + t)/2t < 11.5\varepsilon$.

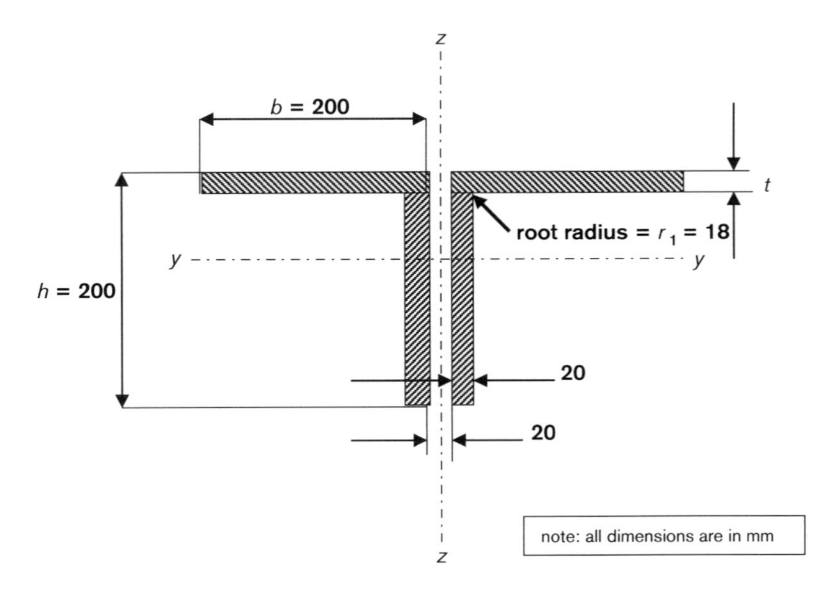

Fig. 5.10. Roof girder: top chord cross-section, two angles 200 × 200 × 20 back to back

Thus the section satisfies the section classification. Also,

$$\bar{\lambda} = L_{cr}/(i_y \lambda_1) = 600/(6.11 \times 86.4) = 1.13.$$

Referring to curve "b" in Fig. 6.4 of Eurocode 3, Part 1-1, $\chi = 0.52$. Therefore

$$N_{b,Rd} = 0.52 \times 15\,300 \times 275/10^3 = 2188 \text{ kN} > N_{Ed} \text{ (1024 kN)}$$

and $N_{Ed}/N_{b,Rd} = 1024/2188 = 0.47 < 1.0$. OK.

Therefore we adopt two angles 200 × 200 × 20 back to back with a 15 mm gap between the vertical faces for the entire top chord.

5.4.8.2 Bottom chord (3-17)

Maximum ultimate design force = 1171 kN (tension).

Try two angles 200 × 200 × 16: $A_g = 124$ cm^2. Deduct two 26 mm holes:

$$A_{net} = 124 - 2 \times 2.6 \times 1.6 = 115.7 \text{ cm}^2,$$

$$N_{t,Rd} = 115.7 \times 275/10 = 3181 \text{ kN} > N_{Ed} \text{ (1171 kN)}.$$

Therefore we adopt two angles 200 × 200 × 20 back to back with a 15 mm gap between the vertical faces for the entire bottom chord. (The thickness of the angles has been increased to satisfy the section classification when the member is subjected to reversal of the compressive stress due to wind action on the horizontal wind girder at the roof truss bottom chord level; see Fig. 5.8.)

5.4.8.3 Diagonal (12-16)

Maximum ultimate design force = 604.3 kN (compression).

Try two angles 120 × 120 × 12: $A_g = 56.2\ \text{cm}^2$, $iy = 3.66$, $L_{cr} = 390$ cm, $\lambda_1 = 86.4$ and

$$\bar{\lambda} = L_{cr}/(iy\lambda_1) = 390/(3.66 \times 86.4) = 1.23.$$

Referring to curve "b" in Fig. 6.4 of Eurocode 3, Part 1-1, $\chi = 0.46$. Therefore $N_{b,Rd} = 0.46 \times 56.2 \times 275/10 = 711\ \text{kN} > N_{Ed}$ (604.3 kN). OK.

Therefore we adopt two angles 120 × 120 × 12 back to back with a 15 mm gap between the vertical faces for all diagonals.

5.4.8.4 Verticals

We adopt two angles 100 × 100 × 12 back to back (see Fig. 5.11, showing the forces in and sizes of members).

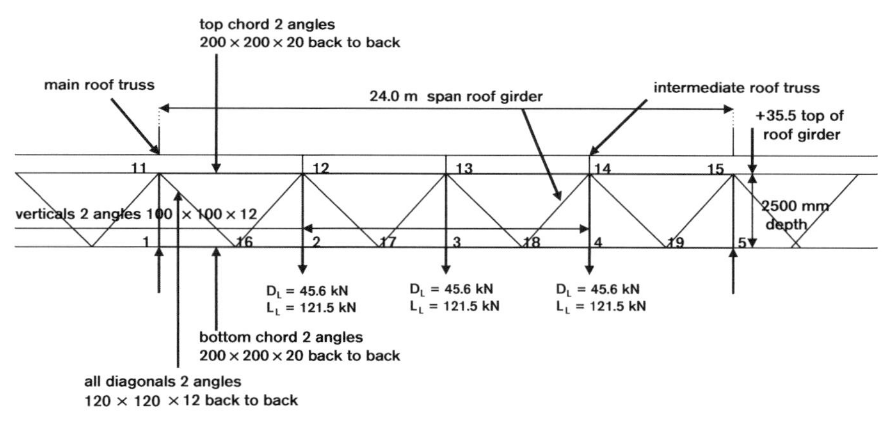

Notes
C indicates compression.
T indicates tension.

Member	Location	Unfactored forces DL (kN)	Unfactored forces DL (kN)	Ultimate forces (kN)	Sizes 2 angles b × h × t
(11–12)	top chord	82.1C	218.7C	2 × 100 × 100 × 10	200 × 200 × 20
(12–13)	top chord	191.5C	510.4C	1024C	200 × 200 × 20
(1–16)	bottom chord	0	0	0	0
(2–16)	bottom chord	164.2T	437.4T	878T	200 × 200 × 20
(2–17)	bottom chord	164.2T	437.4T	878T	200 × 200 × 20
(3–17)	bottom chord	219.0T	583.4T	1171T	200 × 200 × 20
(11–16)	diagonal	106.7T	284.3T	570T	120 × 120 × 12
(12–16)	diagonal	106.7C	284.3C	570C	120 × 120 × 12
(12–17)	diagonal	35.6T	94.8T	190T	120 × 120 × 12
(13–17)	diagonal	35.6C	94.8C	190C	120 × 120 × 12
(1–11)	vertical	68.4C	182.2C	366C	100 × 100 × 12
(5–12)	vertical	45.6T	121.5T	244T	100 × 100 × 12

Fig. 5.11. Forces in and sizes of latticed roof member girders

5.5 Design of intermediate columns (members subjected to bending and thrust)

5.5.1 Design considerations

The intermediate columns, spaced at 6 m between the main stanchions, are assumed to be fixed at the base and simply supported at roof girder level. They are to be designed to resist wind loads on the side wall and the dead load from the side wall.

Vertical span from ground level to underside of roof girder = H = 33.0 m.

Spacing of columns = 6.0 m

(see Figs 5.7 and 5.8).

5.5.2 Functions

The intermediate columns have been provided to serve the following functions:

- To support the side rails at 2 m spacing, which are continuous over the columns.
- To obtain an economical size of the side rails with a 6.0 m span, continuous over the columns.
- To transfer the wind loads at roof bracing level.

5.5.3 Loadings

5.5.3.1 Dead loads

Dead load/m^2 of wall (previously calculated) = 0.138 kN/m^2.

Weight of side rails (with channel 150 × 89 × 23.84 kg/m) spaced at 2.0 m = 23.84/2 = 0.12 kN/m^2.

Total = 0.258 kN/m^2.

Therefore

dead load/m height of column (spacing = 6.0 m) = 6 × 0.258 = 1.548 kN/m height.

Self-weight of column (assuming UB610 × 305 × 238.1 kg/m + 2 × 400 × 20 plate) = 3.64 kN/m height.

Total = 5.19 kN/m height.

Therefore

total load on column = G_k = 33 × 5.19 = 171.3 kN;

ultimate load = N_u = 1.35 × 171.3 = 231 kN.

5.5.3.2 Wind loads

Effective wind pressure normal to wall = $q_p(z)$ = 1.4 kN/m^2 (previously calculated).

Therefore the external wind pressure/m height of column for a 6 m column spacing is

w_k = 1.4 × 6 = 8.4 kN/m.

5.5.4 Moments

It is assumed that the base is fixed and the top is freely supported.

Negative moment at base = $w_k H^2/8 = 8.4 \times 33^2/8 = 1143.4$ kN m.

Positive moment at 5/8 span from base = $w_k(9/128)H^2 = 9/128 \times 8.4 \times 33^2$
= 643 kN m.

With a partial safety factor $\gamma_{wk} = 1.5$,

maximum ultimate design moment $M_{Ed} = 1.5 \times 1143.4 = 1715$ kN m;

maximum ultimate shear at base $V_{Ed} = 1.5 \times (5/8) \times w_k H = 1.5 \times 5/8 \times 8.4 \times 33$
= 260 kN;

maximum ultimate thrust $N_{Ed} = 231$ kN

(see Fig. 5.12, showing the deflected shape of the intermediate columns and the BM and SF diagrams).

5.5.5 Design of section, based on Eurocode 3, Part 1-1

From the deflected shape of the member and the BM diagram shown in Fig. 5.12, we find that the outer compression flange, connected to the side rails, is restrained from lateral buckling for a length of 2.0 m, whereas at a certain height above the fixed base the inner flange is subjected to compression due to the moment (as can be seen from the BM diagram). The compression flange is unrestrained at this height, and lateral–torsional buckling of the member occurs, thus reducing the buckling resistance moment of the member. The position where the member is unrestrained may be assumed to be the point of contraflexure in the BM diagram, and at a height equal to one-quarter of the height from the fixed base of the member (i.e. 1/4 × 33 m, say 8.3 m).

The following sequence of operations was performed in the design of this member.

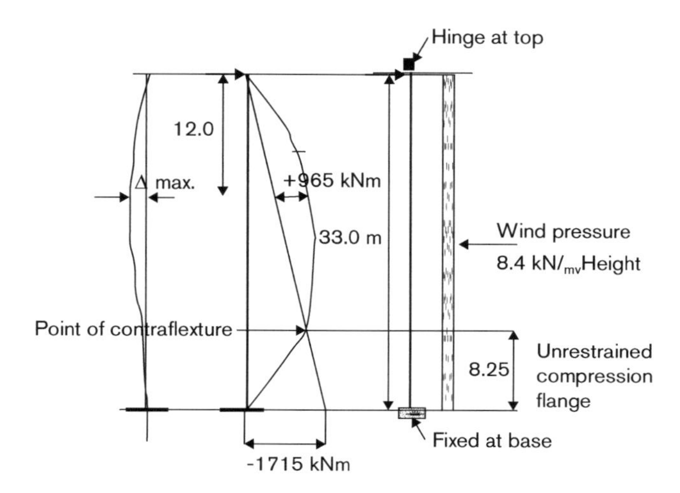

Fig. 5.12. BM diagram and deflected shape of intermediate columns

5.5.5.1 Step 1: Selection of section and obtaining the properties of the section

Try a section UB 914 × 419 × 343 kg/m: $A_g = 437$ cm^2, $W_y = 15\,500$ cm^3, $i_y = 37.8$ cm, $W_z = 2890$ cm^3, $i_z = 9.46$ cm. We assume grade S 275 steel.

Thickness of flange = $t_f = 32$ m.

Thickness of web = $t_w = 19.4$ mm.

Width of flange = $b = 418.5$ mm.

Clear depth of web = $d = 799.6$ mm.

Depth of section = $h = 911.8$ mm.

Depth of web = $h_w = 847.8$ mm.

5.5.5.2 Step 2: Section classification

Before designing the section, we have to classify the section into one of the following classes:

- *Class 1:* cross-section with plastic hinge rotation capacity.
- *Class 2:* cross-section with plastic moment capacity, but limited rotation capacity.
- *Class 3:* cross-section in which the stress in the extreme compression fibre can reach the design strength, but a plastic moment capacity cannot be developed.
- *Class 4, slender:* cross-section in which we have to have a special allowance owing to the effects of local buckling.

Flange

Stress factor = $\varepsilon = (235/275)^{0.5} = 0.92$.

Outstand of flange = $c = (b - t_w - 2r) = (418.5 - 19.4 - 2 \times 24.1)/2 = 175.5$ mm.

Ratio $c/t_f = 175.5/32$ (average) = 5.5, and $9\varepsilon = 9 \times 0.92 = 8.3$.

For class 1 classification, the limiting value of $c/t_f \leq 9\varepsilon$; and we have $5.5 \leq 8.3$.

So, the flange satisfies the condition for class 1 classification. <u>OK</u>

Web

Referring to Table 5.2 (sheet 1) of Eurocode 3, Part 1-1, the web is subject to bending and compression. Assuming that the depth ratio α = (depth of web in compression)/(depth of web in tension) ≤ 0.5,

$d/t_w \leq 36\varepsilon/\alpha$

Ratio $d/t_w = 799.6/19.4 = 41.2$, and ratio $36\varepsilon/\alpha = 36 \times 0.92/0/5 = 66.2 > 41.2$.

So, the web satisfies the conditions for class 1 section classification. Therefore the rolled I section is classified as class 1.

5.5.5.3 Step 3: Checking the shear capacity of the section

The shear force (V_{Ed}) should not be greater than the design plastic shear resistance ($V_{pl,rd}$) of the section. In our case, in the absence of torsion, the design plastic shear resistance is

$$V_{pl,rd} = A(f_y/\sqrt{3})/\gamma_{Mo} \qquad (6.18)$$

where A = shear area

$= A - 2bt_f + (t_w + 2r)t_f = 43\ 700 - 2 \times 418.5 \times 32 + (19.4 + 2 \times 24.1) \times 32$

$= 190.8\ cm^2$ but not less than $h_w t_w = 84.78 \times 1.94 = 164.5\ cm^2$, and

$V_{pl,Rd} = 190.8 \times 10^2 \times (275/\sqrt{3})/10^3 = 3029$ kN.

Ultimate shear force at base = 260 kN << $V_{pl,Rd}$ (3029 kN). Satisfactory

5.5.5.4 Step 4: Checking for shear buckling

If the ratio h_w/t_w exceeds 72ε for a rolled section, then the web should be checked for shear buckling. In our case, $d/t_w = 41.2$ and $36\varepsilon/\alpha = 66.2 > h_w/t_w$. Therefore the web need not be checked for shear buckling.

5.5.5.5 Step 5: Checking the moment capacity of the section

Maximum ultimate design moment at base $M_{Ed} = 1715$ kN m.

Maximum ultimate design shear at base $V_{Ed} = 260$ kN.

Maximum ultimate design thrust at base $N_{Ed} = 231$ kN.

Referring to Clause 6.2.10, where bending, shear and axial forces act simultaneously on a structural member, the moment capacity should be calculated in the following way:

- *When the web is not susceptible to buckling.* When the web-to-depth ratio $h_w/t_w \leq 72\varepsilon$ for class 1 classification, it should be assumed that the web is not susceptible to buckling, and the moment capacity should be calculated from the equation $M_{ypl,rd} = f_y W_y$, provided the design shear force $V_{Ed} < (1/2)V_{pl,Rd}$ and $N_{Ed} < (1/4)N_{pl,Rd}$. In our case, $h_w/t_w < 72\varepsilon$. So, the web is not susceptible to buckling.
- *When the design ultimate shear force* $V_{Ed} < (1/2)V_{pl,Rd}$. In our case, $V_{Ed} = 260$ kN, and $(1/2)V_{pl,Rd} = 1/2 \times 3029$ kN = 1515 kN.
- *When the design ultimate thrust* $N_{Ed} < (1/4)N_{pl,Rd}$. In our case, $N_{Ed} = 231$ kN, and $(1/4)N_{pl,Rd} = (1/4)Af_y = 0.25 \times 190.8 \times 10^2 \times 275/10^3 = 1312$ kN >> 231 kN.

Therefore the plastic moment capacity is

$M_{ypl,Rd} = 275 \times 15\ 500 \times 10^3/10^6 = 4263$ kN m >> 1715 kN m. Satisfactory

5.5.5.6 Step 6: Buckling resistance to compression

Referring to Clause 6.3.1, the buckling resistance to compression is

$$N_{b,Rd} = \chi A f_y/\gamma_{M1} \qquad (6.47)$$

$$\bar{\lambda} = L_{cr}/(i\lambda_1) \qquad (6.50)$$

$L_{cr1} = 0.85L = 0.85 \times 33 = 28$ m (assuming base fixed and top hinged)

(The factor of 0.85 has been taken because the bottom is assumed fixed and the top hinged.)

$\lambda_1 = 93.9\varepsilon = 93.9 \times 0.92 = 86.4$

$iy = 37.8$ cm

$\bar{\lambda} = 28 \times 10^3/(37.8 \times 10 \times 86.4) = 0.86$

Referring to Table 6.2 of Eurocode 3, Part 1-1, for a rolled section, we follow curve "a" in Fig. 6.4 of that Eurocode and find that the reduction factor $\chi = 0.75$. Therefore

$$N_{b,Rd} = 0.75 \times 43\,700 \times 275/10^3 = 9013 \text{ kN} >> 231 \text{ kN} \qquad \underline{\text{OK}}$$

$$\text{Also, } N_{Ed}/N_{b,Rd} = 231/9013 = 0.03 << 1.0$$

5.5.5.7 Step 7: Buckling resistance moment

As discussed previously, the unstrained height of the inner compression flange from the base is 1/4 × 33 = 8.3 m. The buckling resistance moment is calculated in the following way.

$$\text{Design buckling resistance moment about major axis} = M_{b,Rd} = \chi_{LT} \times W_{pl,y} f_y / \gamma_{M1} \qquad (6.55)$$

where

$$\chi_{LT} = 1/(\Phi_{LT} + (\Phi_{LT}^2 - \bar{\lambda}_{LT}^2)^{0.5}) \qquad (5.56)$$

Here, $\Phi_{LT} = 0.5[1 + \alpha_{LT}(\bar{\lambda}_{LT} - 0.2) + \bar{\lambda}_{LT}^2]$ and $\bar{\lambda}_{LT} = \sqrt{[(W_y f_y)/M_{cr}]}$, where M_{cr} is the elastic critical moment for lateral–torsional buckling. It is not specified in the code how to calculate the value of M_{cr}. So, referring to Clause 6.3.2.4, we shall adopt a simplified method.

Members with discrete lateral restraint to the compression flange are not susceptible to lateral–torsional buckling if the length L_c between restraints or the resulting slenderness $\bar{\lambda}_f$ of the equivalent compression flange satisfies

$$\bar{\lambda}_f = (k_c L_c)/(i_f{,}z \lambda_1) \leq \bar{\lambda}_{c0}(M_{c,Rd}/M_{y,Ed}) \qquad (6.59)$$

where $M_{y,Ed}$ is the maximum design value of the bending moment within the restraint spacing, equal to 1715 kN m;

$$M_{c,Rd} = W_y f_y / \gamma_{M1} = 15\,500 \times 275/10^3 = 4263 \text{ kN m;}$$

W_y is the section modulus of the section about the y–y axis, equal to 15 500 cm^3; k_c is the slenderness correction factor (from Table 6.6 of Eurocode 3, Part 1-1, $k_c = 0.91$); L_c is the unrestrained height of the compression flange for lateral–torsional buckling, equal to 830 cm; $i_{f,z}$ is the radius of gyration of the equivalent compression flange, equal to 9.46 cm; and $\bar{\lambda}_{c0}$ is the slenderness limit of the equivalent compression flange.

Referring to Clause 6.3.2.3,

$$\bar{\lambda}_{c0} = \bar{\lambda}_{LT,0} + 0.1 = 0.4 \text{ (recommended)} + 0.1 = 0.5$$

$$\lambda_1 = 93.9\varepsilon = 93.9 \times 0.92 = 86.4.$$

Therefore

$$\bar{\lambda}_f = (k_c L_c)/(i_{f,z} \lambda_1) = (0.91 \times 830)/(8.96 \times 86.4) = 0.98$$

$$\text{and } \bar{\lambda}_{c0}(M_{c,Rd}/M_{y,Ed}) = 0.5 \times 4263/1715 = 1.24 > 0.98.$$

So, no reduction of buckling resistance moment needs to be considered. Therefore the design buckling resistance moment $M_{b,Rd} = W_y f_y / \gamma_{M1} = 4263$ kN m, and $N_{Ed}/N_{b,Rd} + M_{Ed}/M_{bRd} = 0.03 + 1715/4263 = 0.03 + 0.40 = 0.43 < 1$.

Therefore, we adopt the section UB914 × 419 × 343 kg/m.

5.5.5.8 Step 8: Checking deflection

The column is assumed to be fixed at the base and freely supported at the top.

Height of column = 33 m.

Unfactored wind load/m height = w_k = 8.4 kN/m.

Section adopted = UB914 × 419 × 343 kg/m, Ix = 626 000 cm^4, E = 21 000 kN/cm^2.

Maximum deflection = $\Delta_{max} = w_k h^4/(185EI)$

$= 8.4 \times 33 \times 33^3 \times 100^3/(185 \times 21\,000 \times 626\,000) = 4.1$ cm.

Permissible deflection = $\Delta_p = H/360 = 33 \times 100/360 = 9.2$ cm $> \Delta_{act}$. Satisfactory

5.6 Design of horizontal wind bracing system for roof (members subjected to compression and tension)

5.6.1 Design considerations

The horizontal wind bracing system consists of lattice girders (of warren type) formed by the bottom chord members of the vertical roof girders and the other chord members at the bottom chord level of the roof trusses. Two sets of wind girders, one along column line A and another one along column line E, are to be provided (see Fig. 1.2). The chord members are 6 m apart, and bracings connect them to form a horizontal roof wind girder (see Fig. 5.8). The girder spans 24 m between the main roof trusses. The depth of the girder is 2.25 m.

5.6.2 Functions

The wind girder on each column line, either A or E, is assumed to take the full horizontal wind reactions from the intermediate columns as point loads, and forces are induced in the individual members owing to the loadings.

5.6.3 Loadings (wind loads)

From previous calculations,

effective external wind pressure $q_p(z) = 1.4$ kN/m^2.

External wind pressure/m height of intermediate column = $w_k = 1.4 \times 6 = 8.4$ kN.

The intermediate column is assumed to be fixed at the base and simply supported at the top. Therefore the reaction at the top, $R_t = W_k$ at a node, is equal to

$(3/8)w_k H = 3/8 \times 8.4 \times 33 = 104$ kN (unfactored).

5.6.4 Forces in members of braced girder

The braced girder (1-6-10-5) shown in Fig. 5.8 is loaded with point wind loads of $W_k = 104$ kN (unfactored) at the nodal points 2, 3 and 4. It is assumed that all three point loads are taken by the braced girder. To calculate the forces in the members, we first find the reactions at the support:

$R_l = 1.5 \times 104$ kN $= 156$ kN $= R_r$

We then calculate the forces as follows.

5.6.4.1 Forces in members (unfactored)

Consider node 1.

Length of diagonal (1-7) = $(2.25^2 + 6^2)^{0.5}$ = 6.41 m.

$\sum V = 0$.

Therefore

force in member (1-7) × 2.25/6.41 = member (1-6) = R_1 = 156;

force in member (1-7) = 156 × 6.41/2.25 = 444.4 kN (tension).

$\sum H = 0$.

Therefore

(1-2) = (1-7) × 6/6.41 = 444.4 × 6/6.41 = 417 kN (compression).

Next, consider node 7.

$\sum V = 0$.

Therefore

(2-7) + (3-7) × 2.25/6.41 − (1-7) × 2.25/6.41) = 0;

(3-7) = [(444.4 × 2.25/6.41 − 104)] × 6.41/2.25 = 148.8 kN (compression).

$\sum H = 0$.

Therefore

(7-8) = (1-7) × 6/6.41 + (3-7) × 6/6.41 = (444.4 + 148.8) × 6/6.41) = 555.3 kN (tension).

5.6.4.2 Ultimate design forces in members

With a partial safety factor γ_{wk} = 1.5 (due to wind), we have the following ultimate design forces in the members.

In chord member (1-2), force = 1.5 × 416 = 624 kN (compression).

In chord member (2-3), force = 624 kN (compression).

In chord member (6-7), force = 0.0.

In chord member (7-8), force = 1.5 × 555.3 = 833 kN (tension).

In diagonal member (1-7), force = 1.5 × 444.4 = 667 kN (tension).

In diagonal member (3-7), force = 1.5 × 148.8 = 223 kN (compression).

In addition, the chord members (1-2), (2-3), (3-4) and (4-5), acting as bottom chord members of the vertical roof girder, have ultimate tension forces due to the vertical loads (DL + LL) (see Fig. 5.11).

In case 1, when WL is acting with DL only, the load combination to be considered is $1.5 \times \text{WL} + 1.0 \times \text{DL}$. Therefore, in chord member (1-2), the ultimate design force is $624 - 437 = 187$ kN (compression). In chord member (2-3), the ultimate design force is $624 - 583 = 41$ (compression).

In case 2, when WL is acting simultaneously with (DL + LL), the partial safety factor is taken equal to $0.9 \times 1.5 = 0.9$. Therefore, in chord member (1-2), the ultimate design force is $0.9 \times 624 - 878 = -316$ kN (tension), and in chord member (2-3), $0.9 \times 624 - 1171 = -609$ kN (tension) (see Fig. 5.11).

5.6.5 Design of section of members

5.6.5.1 Chord member (7-8)

Ultimate design force = −833 kN (tension).

For the initial sizing of the section, we try two angles $120 \times 120 \times 12$: $A_g = 55$ cm^2. Using two 26 mm holes,

$$A_{net} = 55 - 2 \times 2.6 \times 1.2 = 48.76 \text{ cm}^2.$$

For the design of the section,

$$N_{pl,Rd} = A_{net} f_y / \gamma_{Mo} = 48.76 \times 100 \times 275/10^3 = 1340 \text{ kN} > N_{Ed} \text{ (833 kN)}$$ Satisfactory

Therefore we adopt two angles $120 \times 120 \times 12$ back to back with a 12 mm gap between the vertical faces.

5.6.5.2 Chord member (2-3)

Ultimate design force = 187 kN (compression) when WL and DL are acting, no LL.

Ultimate design force = 1171 kN (tension) when DL + LL are acting, no WL.

Firstly, we design the member as a compression member.

For the initial sizing of the section, we try two angles $200 \times 200 \times 20$: $A_g = 153$ cm^2, $iy = 6.11$ cm, $l = 600$ cm.

For the section classification,

$$h/t = 200/20 = 10,$$

$$15\varepsilon = 15 \times 0.92 = 13.8,$$

$$(h + b)/2t = 400/40 = 10,$$

$$11.5\varepsilon = 11.5 \times 0.92 = 10.6.$$

Referring to Table 5.2 (sheet 3) of Eurocode 3, Part 1-1, for class 3 classification, $h/t \leq 15\varepsilon$ and $(h + b)/2t \leq 11.5\varepsilon$. In our case, the section satisfies both conditions.

For the buckling resistance in compression,

$$N_{b,Rd} = \chi A f_y / \gamma_{M1} \tag{6.47}$$

$$\bar{\lambda} = L_{cr}/(iy\lambda_1) = 600/(6.11 \times 93.9 \times 0.92) = 1.14$$

where $\lambda_1 = 93.9\varepsilon$. Referring to Table 6.2 of Eurocode 3, Part 1-1, for angles, we follow the buckling curve "b" in Fig. 6.4 of that Eurocode, and find $\chi = 0.5$. Therefore

$N_{b,Rd} = 0.5 \times 153 \times 100 \times 275/10^3 = 2100$ kN >> N_{Ed} (187 kN). <u>OK</u>

Now, we design the member as a tension member.

$A_{net} = 153 - 2 \times 2.6 \times 2 = 142.6$ cm^2.

Therefore

$N_{pl,Rd} = A_{net}f_y/\gamma_{Mo} = 142.6 \times 275/10 = 3922$ kN >> N_{Ed} (1171 kN) <u>OK</u>

Therefore we adopt two angles 200 × 2000 × 20 back to back with a 15 mm gap between the vertical faces.

5.6.5.3 Diagonal members (3-7) and (3-9)

Ultimate design force = 667 kN (tension).

Ultimate design force = 223 kN (compression).

Firstly, we design the member as a compression member.

For the initial sizing of the section, we try two angles 120 × 120 × 12 $A = 55$ cm^2: $A_{net} = 55 - 2 \times 26 \times 1.2 = 48.8$ cm^2, $iy = 3.65$ cm, $L_{cr} = 641$ cm.

For the section classification,

$h/t = 120/12 = 10$, $15\varepsilon = 15 \times 0.92 = 13.8$,

$(h + b)/2t = 240/24 = 10$, $11.5 \times \varepsilon = 11.5 \times 0.92 = 10.58$.

Referring to Table 5.2 of Eurocode 3, Part 1-1, to satisfy the conditions for class 3 section classification, $h/t \leq 15\varepsilon$ and $(h + b)/2t \leq 11.5t$, and we have $10 \leq 13.5$ and $10 \leq 10.58$. So, the section satisfies the section classification.

For the buckling resistance in compression,

$N_{b,Rd} = \chi A f_y/\gamma_{M1}$

$\bar{\lambda} = L_{cr}/(iy\lambda_1) = 641/(3.65 \times 93.9 \times 0.92) = 2.03$.

Referring to Table 6.2 of Eurocode 3, Part 1-1, for angles, we follow the buckling curve "b" in Fig. 6.4 of that Eurocode. With $\bar{\lambda} = 2.03$, $\chi = 0.2$. Therefore

$N_{b,Rd} = 0.2 \times 55 \times 275/10 = 303$ kN > 223 kN <u>OK</u>

Now, we design the member as a tension member.

$N_{pl,Rd} = A_{net}f_y/\gamma_{Mo} = 48.8 \times 275/10 = 1342$ kN > 667 kN <u>OK</u>

Therefore we adopt two angles 120 × 120 × 12 back to back with a 12 mm gap between the vertical faces.

For the diagonals (1-7) and (5-9), we adopt two angles 120 × 120 × 10 back to back.

See Fig. 5.13, showing the forces and member sizes for the horizontal wind girder.

HORIZONTAL LATTICED WIND GIRDER

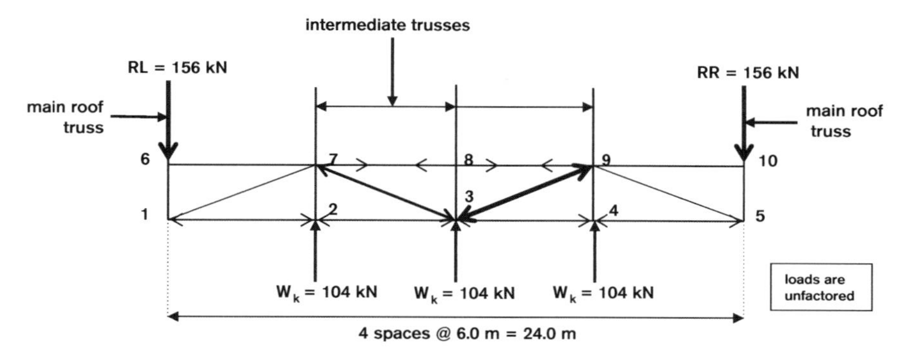

Member	Location	Ultimate force compression (kN)	Ultimate force tension (kN)	Member size 2 angles b × h × t
(1–2)	chord	187	1171	200 × 200 × 20
(2–3)	chord	41	1171	200 × 200 × 20
(3–4)	chord	41	1171	200 × 200 × 20
(4–5)	chord	87	1171	200 × 200 × 20
(6–7)	chord	0	0	120 × 120 × 12
(9–10)	chord	0	0	120 × 120 × 12
(7–8)	chord	0	833	120 × 120 × 12
(8–9)	chord	0	833	120 × 120 × 12
(1–7)	diagonal	0	667	120 × 120 × 12
(5–9)	diagonal	0	667	120 × 120 × 12
(3–7)	diagonal	223	0	120 × 120 × 12
(3–9)	diagonal	223	0	120 × 120 × 12

Fig. 5.13. Wind girder: loadings and ultimate forces in members, and sizes of members

References

Eurocode, 2002a. BS EN 1991-1-1: 2002, Actions on structures. General actions. Densities, self-weight, imposed loads for buildings.

Eurocode, 2002b. BS EN 1990: 2002(E), Basis of structural design.

Eurocode, 2005a. BS EN 1991-1-4: 2005, Actions on structures. General actions. Wind actions.

Eurocode, 2005b. BS EN 1993-1-1: 2005, Eurocode 3. Design of steel structures. General rules and rules for buildings.

Reynolds, C.E. and Steedman, J.C., 1995. *Reinforced Concrete Designer's Handbook*, Spon, London.

CHAPTER 6

Case Study I: Analysis and Design of Structure of Melting Shop and Finishing Mill Building

6.1 Design considerations

The melting shop and finishing mill building comprise a framing system consisting of a series of four frames with stanchions and roof trusses. The trusses are assumed to be hinge-connected to the top of the stanchions, and the bases of stanchions are assumed to be fixed (see Figs. 1.1–1.5). The frames are subjected to vertical dead, live and crane loads, and also horizontal transverse crane surge and wind loads. The forces due to the resultant wind pressures on the structure are assumed to be shared by the stanchions in proportion to the stiffness of the stanchions.

6.2 Loadings

6.2.1 Wind loads, based on Eurocode 1, Part 1-4 (Eurocode, 2005a)

6.2.1.1 Wind forces acting from right to left on the structure (see Fig. 6.1)

The external wind pressure normal to a vertical wall face is

$$w_e = q_p(z)c_{pe} = 1.27 \times 1.1 = 1.4 \text{ kN/m}^2$$

where $q_p(z)$ is the peak wind velocity pressure at 47 m height = 1.27 kN/m^2 and c_{pe} is the resultant pressure coefficient on the windward vertical face = 1.1 (see Fig. 2.2(c)). (All of the above values have been calculated previously.) The width of wall on which the wind pressure acts directly on the stanchion is 6.0 m. Therefore

$$\text{wind force/m height of stanchion} = w_{k1} = 1.4 \times 6 = 8.4 \text{ kN/m}.$$

In addition, the reaction from the horizontal roof wind girders on the windward face of the wall acting at the top of the stanchion, P_{h1}, is three times the reaction of each intermediate column,

$$3W_k = 3 \times 104 = 312 \text{ kN acting at 35.5 m from ground level (see Fig. 5.13)}$$

The resultant wind pressure on the projected windward and leeward faces of the storage hopper bay is

$$w_e = q_p(z)c_{pe} = 1.27 \times 1.3 = 1.65 \text{ kN/m}^2.$$

The spacing of the stanchions s is 24 m and the projected height h_2 is 10 m. Therefore

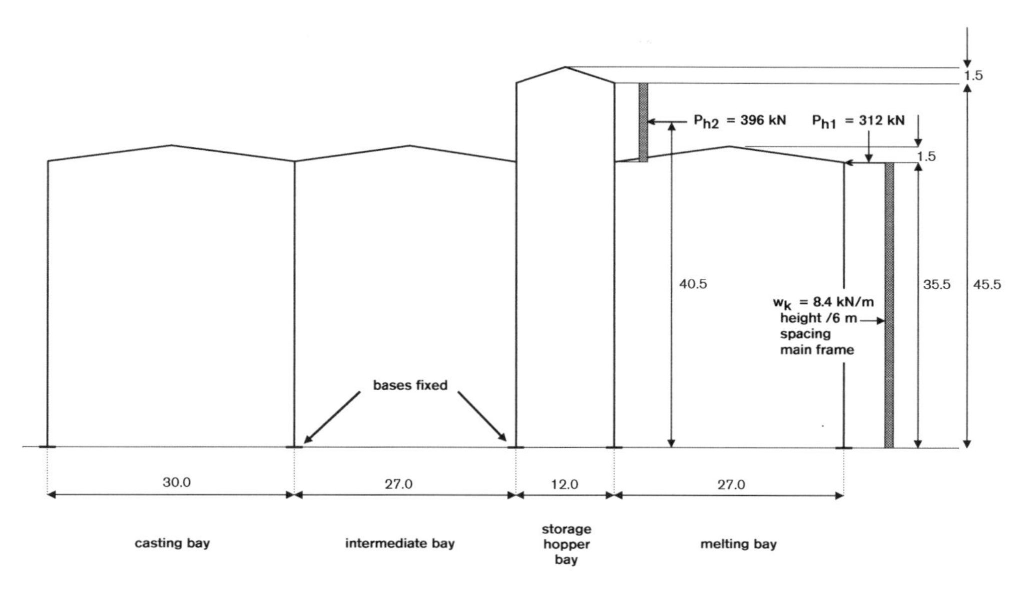

Fig. 6.1. Effective wind forces (wind from right) on the structure

total wind force $= P_{h2} = w_e h_2 s = 1.65 \times 10 \times 24 = 396$ kN acting at 40.5 m from ground level.

6.2.1.2 Wind forces acting from left to right on the frame (see Fig. 6.2)

The resultant external wind pressure normal to the exposed upper portion of a wall face is

$$w_e = q_p(z)c_{pe} = 1.27 \times 1.1 = 1.4 \text{ kN/m}^2$$

where $q_p(z)$ is the peak wind velocity pressure at 47 m height = 1.27 kN/m^2 and c_{pe} is the resultant pressure coefficient on the windward vertical face = 1.1 (see Fig. 2.2(c)). (All of the above values have been calculated previously.) The width of wall on which the wind pressure acts on the stanchion is 24 m. Therefore

wind force/m height of stanchion $= w_{k1} = 1.4 \times 24 = 33.6$ kN/m.

The height of the exposed wall face h_1 is $35.5 - 23.5 = 12$ m. Therefore

total wind force on the exposed wall face $= P_{h1} = w_{k1}h_1 = 33.6 \times 12 = 403$ kN $\rightarrow$

acting at a height of 29.5 m from ground level.

The resultant wind pressure on the projected windward and leeward wall faces of the storage hopper bay is

$$w_e = q_p(z)c_{pe} = 1.27 \times 1.3 = 1.65 \text{ kN/m}^2.$$

The spacing of the stanchions s is 24 m and the projected height h_2 is 10 m. Therefore

total wind force $= P_{h2} = w_e h_2 s = 1.65 \times 10 \times 24 = 396$ kN acting at 40.5 m from ground level.

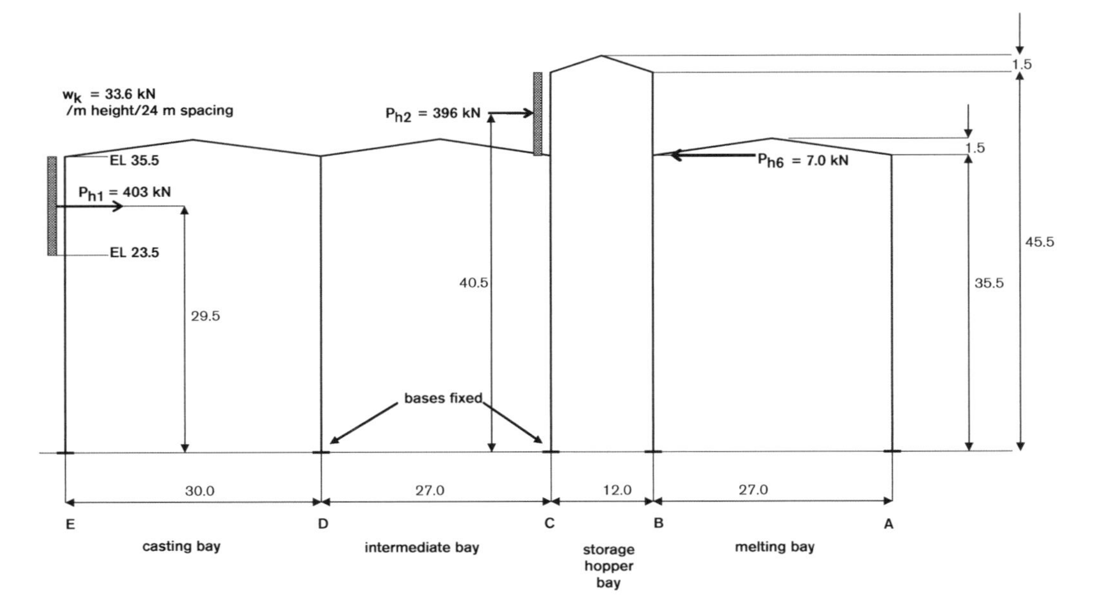

Fig. 6.2. Effective wind forces (wind from left) on the structure

6.2.2 Moment due to wind

6.2.2.1 Moment when wind is blowing from right to left (see Fig. 6.1)

$M_{wl} = w_{k1} \times 35.5^2/2 + P_{h1} \times 35.5 + P_{h2} \times 40.5$

$= 8.4 \times 35.5^2/2 + 312 \times 35.5 + 396 \times 40.5$

$= 32\,407$ kN m.

6.2.2.2 Moment when wind is blowing from left to right (see Fig. 6.2)

$M_{wr} = P_{h1} \times 29.5 + P_{h2} \times 40.5 = 403 \times 29.5 + 396 \times 40.5$

$= 27\,927$ kN m.

6.2.2.3 Distribution of moments among the stanchions

We assume that the total moment due to wind will be shared by the five stanchions in proportion to the stiffness of the members. The stanchions are assumed to be cantilevers. The deflections at the free ends will vary as Ph^3/I (where h is the cantilever length, I is the moment of inertia of each column and P is the horizontal load). Since the heights of the stanchions vary, the moment shared by each stanchion will vary directly with the moment of inertia and inversely with the cube of the height of each stanchion in order for each stanchion to deflect horizontally by the same amount.

In our case, the stanchions along lines A, D and E are 35.5 m and the stanchions along lines B and D are 45.5 m in height. The proportions of the bending moment to be taken by each stanchion will be

$M_a = M_w(I_a/h_a^3)/(\sum I/h^3)$

$M_b = M_w(I_b/h_b^3)/(\sum I/h^3)$

and so on (where h is the height of each stanchion and I is the moment of inertia of the stanchion).

Now, to calculate the moment of inertia of each stanchion. We assume that each stanchion is made up of the following sections (see Fig. 6.3).

Stanchion A is made of two universal beam sections separated by a distance of 2.5 m:

- Roof leg: UB914 × 305 × 201, $A_g = 256.1$ cm^2
- Crane leg: UB914 × 419 × 388, $A_g = 493.9$ cm^2

Let x_1 be the distance of the centre of gravity from the crane column:

$$\sum A_g x_1 = 256.1 \times 250$$

$$x_1 = 256.18250/(256.1 + 493.9) = 85.4 \text{ cm from crane leg}$$

Therefore

$$I_{ay} = 493.9 \times 85.4^2 + 256.1 \times 164.6^2 = 10\,540\,650 \text{ cm}^4$$

Stanchions B and C are each made of two built-up plate girders separated by a distance of 3.0 m:

- Crane leg: two flange plates 500 × 40 mm + 920 × 20 mm web plate,

$$A_{g1} = 92 \times 2 + 2 \times 50 \times 4 = 584 \text{ cm}^2$$

- Roof leg: two flange plates 500 × 40 mm + 920 × 20 mm web plate,

$$A_{g2} = 584 \text{ cm}^2$$

$$I_{by} = 2 \times 584 \times 150^2 = 26\,280\,000 \text{ cm}^4$$

$$I_{cy} = 26\,280\,000 \text{ cm}^4$$

Stanchion D is made of two built-up columns separated by a distance of 3.0 m:

- Right crane leg: two flange plates 500 × 40 mm + 920 × 20 mm web,

$$A_{g1} = 92 \times 2 + 50 \times 4 \times 2 = 584 \text{ cm}^2$$

- Left crane leg: two flange plates 500 × 30 mm + 940 × 20 mm,

$$A_{g2} = 94 \times 2 + 50 \times 3 \times 2 = 488 \text{ cm}^2$$

$$\sum A_g = 584 + 488 = 1072 \text{ cm}^2$$

Let x_1 be the distance of the centre of gravity from the left leg:

$$x_1 = 584 \times 300/1072 = 163 \text{ cm}$$

$$I_{dy} = 488 \times 163^2 + 584 \times 137^2 = 23\,714\,216 \text{ cm}^4$$

Stanchion E is made of two UB columns separated by a distance of 2 m:

- Crane leg and roof legs: UB 914 × 305 × 201 kg/m

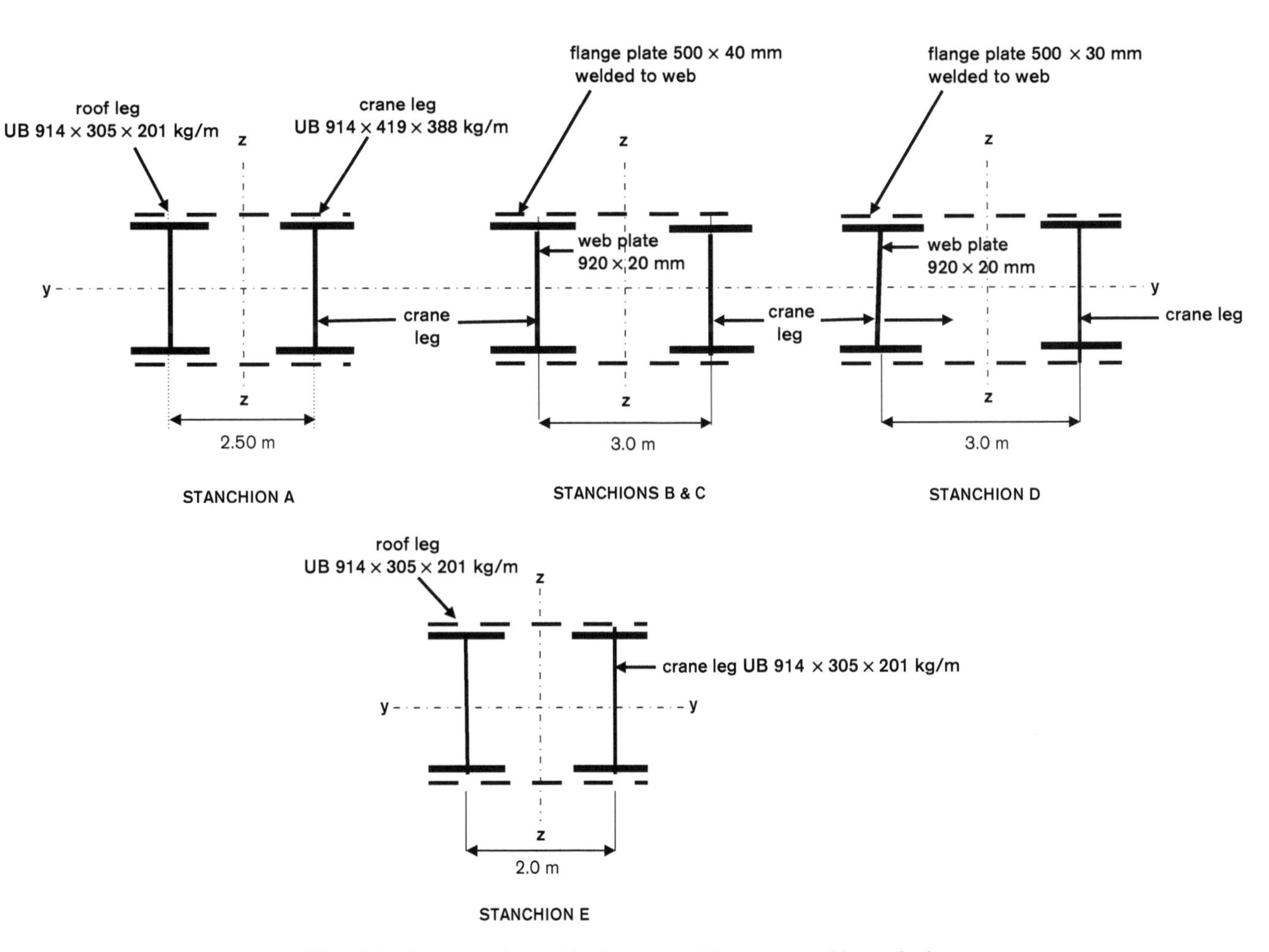

Fig. 6.3. Cross-sections of built-up stanchions assumed in analysis

$A_g = 256.1\ \text{cm}^2$,

$I_{ey} = 2 \times 256.1 \times 100^2 = 5\ 122\ 000\ \text{cm}^4$.

For stanchion A, $I_{ay}/h_a^3 = 10\ 540\ 650/3550^3 = 0.000236$.
For stanchions B and C, $I_{by}/h_b^3 = 26\ 280\ 000/4550^3 = 0.000279$.
For stanchion D, $I_{dy}/h_d^3 = 23\ 714\ 216/3550^3 = 0.000530$.
For stanchion E, $I_{ey}/h_e^3 = 5\ 122\ 000/3550^3 = 0.000114$.

$\sum I/h^3 = 0.000236 + 0.000279 + 0.000279 + 0.00053 + 0.000114 = 0.001438$.

6.2.2.4 Moment shared by stanchions A, B, C, D and E when wind is blowing from right

$M_{al} = \sum M_w I_{ay}/[h_a^3(\sum I/h^3)] = 32\ 407 \times 10\ 540\ 650/[(3550^3) \times 0.001438] = 5310$ kN m.

$M_{bl} = \sum M_w I_{by}/[h_b^3(\sum I/h^3)] = 32\ 407 \times 0.000279/0.001438 = 6288$ kN m;

$M_c = 6288$ kN m.

$M_{dl} = \sum M_w I_{dy}/h_d^3/(\sum I/h^3) = 32\ 407 \times 0.00053/0.001438 = 11945$ kN m.

$M_{el} = \sum M_w I_{ey}/h_e^3/(\sum I/h^3) = 32\ 407 \times 0.000114/0.001438 = 2570$ kN m.

6.2.2.5 Moment shared by stanchions A, B, C, D and E when wind is blowing from left

$M_{ar} = 5310 \times 27\ 932/32\ 407 = 4577$ kN m.

$M_{br} = 6288 \times 0.86 = 5408$ kN m; $M_{cr} = 5408$ kN m.

$M_{dr} = 11\ 945 \times 0.86 = 10\ 273$ kN m.

$M_{er} = 2570 \times 0.86 = 2210$ kN m.

6.3 Design of stanchions in melting bay along line A

See Fig. 6.4. The spacing of stanchions is 24.0 m.

6.3.1 Loadings on crane column

- From maximum vertical crane girder reaction, $V_{max} = 4062$ kN (unfactored).
- From minimum vertical crane girder reaction, $V_{min} = 1357$ kN (unfactored).
- From self-weight reaction of crane girder, 279 kN (unfactored).
- From horizontal crane girder surge, $H_t = 22 \times 5.58 = 122.8$ kN (unfactored).

6.3.2 Loadings on roof column

- From vertical dead load reaction of roof truss directly on column, 44.4 kN (unfactored).
- From vertical live load reaction of roof truss directly on column, 121.5 kN (unfactored).

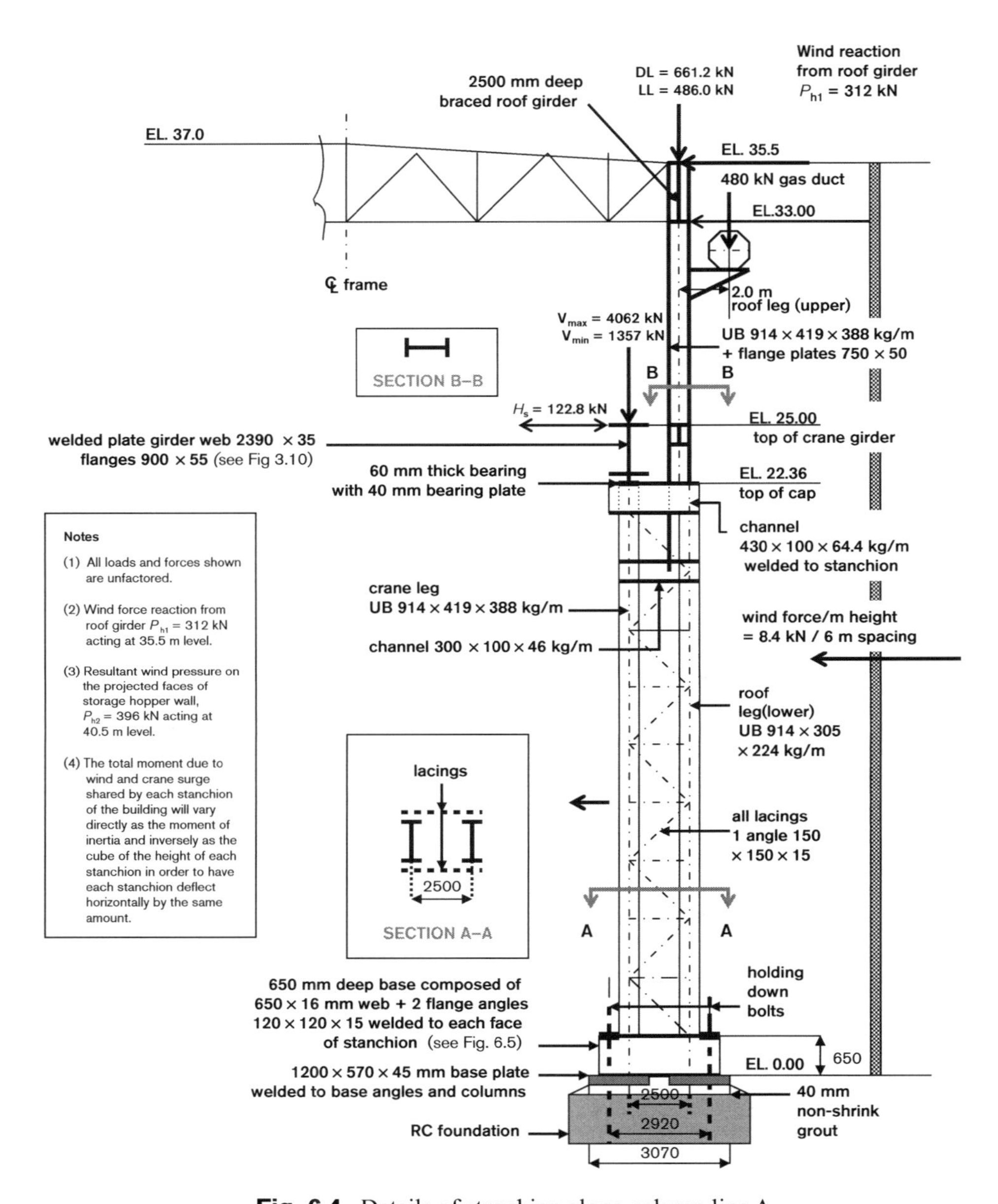

Fig. 6.4. Details of stanchion along column line A

- From vertical dead load reaction of roof girder, 45.6 × 3 = 136.8 kN (unfactored).
- From vertical live load reaction of roof girder, 121.5 × 3 = 364.5 kN (unfactored).
- From vertical dead load reaction of gas duct, 480 kN (unfactored).

$\sum$DL = 44.39 + 136.8 + 480 = 661.2 kN.

$\sum$LL = 121.5 + 364.5 = 486.0 kN.

6.3.3 Moments in stanchion A

- Moment shared by stanchion due to wind blowing from right (previously calculated), $M_{al} = 5310$ kN m (unfactored).
- Moment shared by stanchion due to wind blowing from left (previously calculated), $M_{ar} = 4577$ kN m (unfactored).
- Moment due to crane surge at crane girder level (25.0 m), $122.8 \times 25 = 3070$ kN m (unfactored).

Case 1: when the wind is blowing from the right and the surge is acting from right to left,

maximum moment $\sum M = 5310 + 3070 = 8380$ kN m.

The spacing of the crane and roof columns is 2.5 m (the crane and roof columns form a braced stanchion with a spacing of 2.5 m). Therefore

maximum thrust on each leg due to wind $= \pm 5310/2.5 = \pm 2124$ kN,

maximum thrust on each leg due to crane surge $= \pm 3070/2.5 = \pm 1228$ kN.

Case 2: when the wind is blowing from the left and the surge is acting from left to right,

maximum thrust on each leg due to wind $= \pm 4577/2.5 = \pm 1831$ kN,

maximum thrust on each leg due to crane surge $= \pm 3070/2.5 = \pm 1228$ kN.

6.3.4 Design of sections of stanchions, based on Eurocode 3, Part 1-1 (Eurocode, 2005b) (see Fig. 6.4)

6.3.4.1 Crane column

We consider case 1, where there is a wind load and a fully loaded crane, with a surge acting from right to left simultaneously with (DL + LL). Referring to Table A1.2(B) of BS EN 1990: 2002(E) (Eurocode, 2002b), the partial safety factor for the permanent action (DL) is $\gamma_{Gj} = 1.35$, the partial safety factor for the leading variable action (LL) is $\gamma_{Qk1} = 1.5$ and the partial factor for the accompanying variable action (WL) is $\gamma_{Wk1}\psi_0 = 1.5 \times 0.6 = 0.9$ (for the value of ψ, refer to Table A1.1 of that Eurocode). Therefore the maximum thrust on the crane column is

$N_{Edmax} = 1.5 \times [1228$ (moment due to crane surge) $+ 0.9 \times 2124$ (moment due to wind)

$+ 1.35 \times [279$ (self-weight of crane girder)

$+ 3.88 \times 25$ (assuming UB914 × 419 × 388 kg/m column)] $+ 1.5 \times 4062$ (crane live reaction)

$- 480 \times 2/2.5$ (tension due to gas duct weight)] $= 9970$ kN.

Try a section UB914 × 419 × 388 kg/m: $A_g = 493.9$ cm^2, $i_y = 38.1$ cm, $i_z = 9.27$ cm. We assume $l_y = 0.85 \times$ height up to cap level of stanchion $= 0.85 \times 22.5$ m $= 19.13$ m $=$ 1913 cm, $l_z =$ spacing of horizontal battens in stanchion $= 2.5$ m $= 250$ cm, and $\lambda_1 = 93.9\varepsilon = 93 \times 0.92 = 86.4$. For buckling about the major axis y–y,

$$\bar{\lambda} = L_{cr}/(i_y\lambda_1) = 1913/(38.1 \times 86.4) = 0.58.$$

Referring to Table 6.2 of Eurocode 3, Part 1-1 (see Appendix B), for buckling of a rolled section with $t_f \leq 40$ mm about the y–y axis, we follow the buckling curve "a" in Fig. 6.4 of that Eurocode. With $\bar{\lambda} = 0.58$, the reduction factor $\chi = 0.88$.

For buckling about the minor axis z–z,

$$\bar{\lambda} = 250/(i_z\lambda_1) = 250/(9.27 \times 86.4) = 0.31$$

We follow the curve "b" in Fig. 6.4 of Eurocode 3, Part 1-1, and find that $\chi = 0.95$. So, the y–y axis gives a lower value. We adopt $\chi = 0.88$. Therefore

$N_{b,Rd}$ = design buckling resistance of member in compression

$= \chi A f_y/\gamma_{M1} = 0.88 \times 493.9 \times 275/10 = 11\ 952\ \text{kN} > N_{Ed}\ (9970\ \text{kN})$ Satisfactory

Therefore, we adopt UB914 × 419 × 388 kg/m.

6.3.4.2 Roof column leg (lower portion)

We consider case 1: we assume that the fully loaded crane is in operation + (DL + LL), with the wind from the left.

Vertical (DL + LL) = 661.2 + 486.0 = 1147.2 kN.

Moment due to crane surge at crane girder level (25.0 m) = 122.8 × 25 = 3070 kN m.

Thrust in roof column due to crane surge = 3070/2.5 = 1228 kN.

Moment due to eccentricity of gas duct for force of 480 kN = 480 × 2 = 960 kN m.

Thrust in roof column due to gas duct eccentricity = 960/2.5 = 384 kN.

Moment due to wind blowing from left = 4577 kN m.

Thrust due to wind = 4577/2.5 = 1831 kN.

Therefore the ultimate thrust in the roof leg is

1.35 × 661.2 (DL) + 1.35 × 384 (gas duct)

+ 1.35 × 2.01 × 22.5 (assuming UB914 × 305 × 201 kg/m)

+ 1.5 × 1831(WL) + 0.9 × 1228 (crane surge) + 0.9 × 486 (LL) = 4656 kN.

Try a section UB914 × 305 × 201 kg/m: $A_g = 256.1\ \text{cm}^2$, $i_y = 35.6$, $l_y = 85 \times 22.5 = 1913$ cm.

$$\lambda_1 = 93.9\varepsilon = 93.9 \times 0.92 = 86.4,$$

$$\bar{\lambda} = L_{cr}/(iy\lambda_1) = 1913/(35.6 \times 86.4) = 0.62.$$

Referring to Table 6.2 of Eurocode 3, Part 1-1, for the y–y axis of a rolled I section, we follow the curve "a" in Fig. 6.4 of that Eurocode. With $\bar{\lambda} = 0.62$, the reduction factor $\chi = 0.88$. Therefore the buckling resistance in compression is

$$N_{b,Rd} = \bar{\lambda} A f_y/\gamma_{M1} = 0.88 \times 256.1 \times 275/10 = 6198\ \text{kN} > N_{Ed}\ (4656\ \text{kN})$$

and $N_{Ed}/N_{b,Rd} = 4656/6198 = 0.75 < 1.0$ Satisfactory

Therefore we adopt UB914 × 305 × 224 kg/m for the lower roof column to facilitate the fabrication of the stanchion, with almost the same depth of the crane leg braced with angle lacings.

6.3.4.3 Roof column leg (upper portion)

Total (DL) = 661.2 kN (from roof) + 480 kN (from gas duct) = 1141.2 kN.

Total (LL) = 486 kN (from roof).

Consider the roof leg to be fixed at 22.5 m from ground level and cantilevered for a height of 35.5 − 22.5 = 13 m.

Case 1: (DL + LL + WL), wind acting from left to right.

Shared moment in stanchion A due to wind = 4577 kN m (previously calculated).

It is assumed that the distribution of the moment is triangular. Therefore

moment at 22.5 m level = 4577 × 13/35.5 = 1676 kN m;

moment due to eccentricity of gas duct of 2.0 m = 480 × 2 = 960 kN m (clockwise).

Total moment clockwise = M = (1676 + 960) kN m = 2636 kN m.

Case 2: (DL + LL + WL), wind acting from right to left.

Moment due to wind shared by the stanchion A at ground level = 5310 kN m (previously calculated).

Assuming a triangular moment distribution,

moment at 22.5 m level M(22.5) = 5310 × 13/35.5 = 1945 kN m (anticlockwise),

moment due to eccentricity of gas duct = 480 × 2 = 960 kN m (clockwise).

Net moment M = 1945 − 960 = 985 kN m << 2636 kN m when wind blowing left to right.

Therefore case 1 will govern the maximum design stress. Since only dead and live loads are acting, the partial safety factors γ_f for the combined dead and live loads and the wind load will be as follows (referring to BS EN 1990: 2002):

for DL, $\gamma_{Gj} = 1.35$;

for LL (leading variable), $\gamma_{Qk,1} = 1.5$;

for WL (accompanying variable), $\gamma_{Wk,1} = 0.6 \times 1.5 = 0.9$.

Therefore the vertical ultimate design load is

$$N_{Ed} = 1.35 \times DL + 1.5 \times LL + 0.9 \times WL = 1.35 \times 1141.2 + 1.5 \times 486 = 2270 \text{ kN}$$

and the maximum ultimate design moment is

$$M_u = M_{Ed} = 1.35 \times 1676 + 1.5 \times 960 = 3707 \text{ kN m}.$$

Try a section UC914 × 419 × 388 kg/m + two flange plates 750 × 50: $A_g = 1244$, $i_z = 17.9$ cm, $W_{pl,y} = 48\ 710$ cm^3.

Flange thickness = t_f = 44 mm (average); web thickness = t_w = 21.4 mm.

We assume an effective height $L_y = 1.5 \times 13 = 19.5$ m and a spacing of side rails $L_z = 2$ m. (A factor of 1.5 has been used as it is assumed that the base at crane level is fixed and, at the top, that the roof truss connection is just hinged and is able to sway as a result of wind loadings on the stanchion.)

To check the direct stress,

$$\lambda_1 = 86.4$$

$$\bar{\lambda}_y = L_y/(i_y\lambda_1) = 19.5 \times 100/(44.7 \times 86.4) = 0.50.$$

Referring to Table 6.2 of Eurocode 3, Part 1-1, for buckling of a rolled I section about the y–y axis, we follow curve "c" in Fig. 6.4 of that Eurocode. With $\bar{\lambda}_y = 0.5$, $\chi = 0.85$. Therefore

$$N_{b,Rd} = \chi A f_y/\gamma_{M1} = 0.85 \times 1244 \times 10^2 \times 275/10^3 = 29\ 079 \text{ kN} >> N_{Ed} \text{ (2270 kN)}.$$

To check for moment capacity,

ultimate shear at base = V_{Ed} = 1.5 × shear developed due to wind

$= 1.5 q_p c_{pe} H \times \text{spacing} = 1.5 \times 1.1 \times 1.27 \times 13 \times 6 = 163$ kN;

plastic shear resistance = $V_{pl,rd} = t_w h_w f_y = 21.4 \times 921 \times 275/10^3 = 5420$ kN,

where t_w is the thickness of web and $h_w = h$ is the depth of the beam. We thus find that the ultimate shear is less than half the plastic shear resistance. So, no reduction of moment resistance needs to be considered. Therefore the moment capacity is

$$M_{y,Rd} = W_y f_y = 48\ 710 \times 103 \times 275/10^6 = 13\ 395 \text{ kN m} > 3707 \text{ kN m}. \qquad \underline{\text{OK}}$$

To check for lateral–torsional buckling,

effective height in y–y direction L_{cry} = 19.5 m.

Referring to Clause 6.3.2.4 of Eurocode 3, Part 1-1,

$$\bar{\lambda}_f = (k_c L_c)/(i_{f,z}\lambda_1) \leq \bar{\lambda}_{c0}(M_{c,Rd}/M_{y,Ed}) \qquad (6.59)$$

where $M_{y,Ed}$ is the maximum design value of the bending moment within the restraint spacing, equal to 3707 kN m;

$$M_{c,Rd} = W_y f_y/\gamma_{M1} = 48\ 710 \times 275/10^3 = 13\ 395 \text{ kN m;}$$

W_y is the section modulus of the section about the y–y axis, equal to 48 710 cm^3; k_c is the slenderness correction factor, equal to 1.0 (see Table 6.6 of Eurocode 3, Part 1-1); L_c is the unrestrained height of the compression flange for lateral–torsional buckling, equal to 1950 cm; $i_{f,z}$ is the radius of gyration of the equivalent compression flange, equal to 11 cm;

and $\bar{\lambda}_{c0}$ is the slenderness limit of the equivalent compression flange. Referring to Clause 6.3.2.3 of Eurocode 3, Part 1-1,

$$\bar{\lambda}_{c0} = \bar{\lambda}_{LT,0} + 0.1 = 0.4 \text{ (recommended)} + 0.1 = 0.5$$

$$\lambda_1 = 93.9\varepsilon = 93.9 \times 0.92 = 86.4.$$

Therefore

$$\bar{\lambda}_f = (k_c L_c)/(i_{f,z}\lambda_1) = (1.0 \times 1950/(17.9 \times 86.4) = 1.26$$

and $\bar{\lambda}_{c0}(M_{c,Rd}/M_{y,Ed}) = 0.5 \times 13\ 395/3707 = 1.8 > 1.26.$

So, no reduction of buckling resistance moment needs to be considered. Therefore

$$N_{Ed}/N_{b,Rd} + M_{Ed}/M_{c,Rd} = 2270/29\ 079 + 3707/13\ 395 = 0.1 + 0.3 = 0.4 < 1.$$

Therefore we adopt UB914 × 419 × 388 kg/m + two flange plates 750 × 50 mm for the upper portion of the stanchion (see Fig. 6.4).

6.3.4.4 Design of lacings

Design considerations

The lacings are designed to act as members to tie together the two components of a stanchion so that the stanchion will behave as a single integral member. The lacings may be flat bars or of angle, channel or T-section and are arranged in a bracing system at angles of 45–60° on both faces of the stanchion. The lacings should be designed to resist the total shear arising from the following items:

- the shear force on the stanchion due to crane surge and wind;
- 2.5% of the total vertical axial load (DL + LL) on the stanchion;
- 2.5% of the thrust generated by the moment caused by wind and crane surge and eccentricity of vertical loads.

Calculation of shear forces in the lacing members

We consider the shear force at the base of the stanchion due to wind, calculated previously.

Wind force acting directly on the stanchion = $W_{k1} = w_{k1}h = 8.4 \times 35.5 = 29.2$ kN.

Wind force reactions from intermediate side columns = $P_{h1} = 312$ kN.

Wind force on projected windward and leeward faces of storage hopper structure = $P_{h2} = 396$ kN.

Unfactored total = 1006.2 kN.

We assume that the total shear is shared equally by the five stanchions. Therefore

shear/stanchion = 1006.2/5 (unfactored) = 201.2 kN.

In addition, the stanchion is subjected to transverse crane surge:

crane surge = H_s (unfactored) = 122.8 kN.

2.5% of the axial force in the stanchion is to be taken as shear, and hence we have the following.

Loading on crane column = V_{max} = 4062 kN;

2.5% of crane column load = 0.025 × 4062 = 102 kN.

Loading on roof column (DL) = 661.2 kN.

Loading from gas duct (DL) = 480 kN.

Total DL = 1141.2 kN.

2.5% of DL = 0.025 × 1141.2 = 29 kN.

Loading on roof column (LL) = 486 kN;

2.5% of LL = 0.025 × 486 = 12 kN.

We calculate the cantilever moment in the stanchion as follows.

Spacing of crane and roof legs = 2.5 m.

Shared moment due to wind (calculated previously) = 5310 kN m (anticlockwise).

Thrust in leg = 5310/2.5 = 2124 kN.

2.5% of 2124, as shear in the stanchion, = 0.025 × 2124 = 53 kN.

Moment due to crane surge at crane girder level = 122.8 × 25 = 3070 kN m (anticlockwise).

Thrust in leg = 3070/2.5 = 1228 kN.

2.5% of 1228, as shear in the stanchion, = 31 kN.

Moment due to eccentric load of gas duct (clockwise) = 480 × 2 = 960 kN m.

Thrust in leg = 960/2.5 = 384 kN.

2.5% of 384, as shear in the stanchion, = 10 kN.

Since (DL + LL + WL + crane load) are all acting simultaneously, the partial safety factor is obtained as follows, using the load combinations in the ULS method.

For DL, γ_{Gj} = 1.35; for LL, γ_{Qk} = 0.6 × 1.5 (accompanying variable);

for WL, γ_{Wk} = 1.5 (leading variable);

for crane LL, γ_{Crk} = 0.6 × 1.5 = 0.9 (accompanying variable).

Ultimate design shear = $1.35G_k + 1.5W_k + 0.9(Q_k + C_{rk})$.

Therefore the ultimate design shear to be resisted by the lacings is

1.35 × (29 + 10) + 1.5 × (201.2 + 53) + 0.9 × (12 + 133) = 564 kN.

Ultimate shear resisted by the lacing on each face = V_v = 564/2 = 282 kN.

Force in a diagonal lacing with 45° inclination = 282 × 1.414 = 399 kN (compression).

Design of section of lacings

Force in the diagonal lacing $N_{Ed} = 399$ kN (compression).

Try one angle 150 × 150 × 15: $A_g = 43.0$ cm^2, $i_{r\,min} = 2.93$ cm and L_{cr} is assumed to be the effective length of the lacing = $L = 1.414 \times 250 = 354$ cm.

$$\lambda_1 = 93.9 \times 0.92 = 86.4$$

$$\bar{\lambda} = L_{cr}/(i\lambda_1) = 354/(2.93 \times 86.4) = 1.4.$$

Referring to Table 6.2 of Eurocode 3, Part 1-1, for an angle cross-section, we follow the buckling curve in Fig. 6.4 of that Eurocode. With $\bar{\lambda} = 1.4$, $\chi = 0.38$. Therefore

$$N_{c,Rd} = \chi A f_y/\gamma_{M1} = 0.38 \times 43 \times 100 \times 275/103 = 449 \text{ kN} > N_{Ed} \text{ (399 kN)}$$

and $N_{Ed}/N_{b,Rd} = 399/449 = 0.89 < 1$ Satisfactory

Therefore we adopt one angle 150 × 150 × 15 for all diagonal and horizontal lacings.

6.3.5 Design of holding-down (anchor) bolts

6.3.5.1 Design considerations

The holding-down bolts should be designed to resist the tension developed owing to the holding of the bases of the stanchions when subjected to tension resulting from the moment at the base. The bolts should have an adequate anchorage length in the foundation concrete to resist the tension. To allow the final adjustment of the base into its exact position, it is normal practice to cast the holding-down bolts in the concrete within a pipe sleeve so that there remains some tolerance of the holding-down bolts in any direction. To resist shear at the base, either the anchor bolts should be designed to take shear in addition to tension or a separate shear angle welded to the anchor bolts should be provided at the top surface of the foundation, embedded in the concrete. The tension capacity of the holding-down bolts is given by $N_{t,Rd} = 0.8 f_y A_{net}$, where f_y is the yield strength of the bolt and A_{net} is the area at the bottom of the threads.

6.3.5.2 Calculations of loads on the stanchion legs (see Fig. 6.5)

Vertical loads on crane leg (maximum)

Reaction from crane girder = $V_{cr\,max} = 4062$ kN (variable action).

Weight of crane girder = 279 kN (permanent action).

Self-weight of crane leg = 3.88 × 22.36 = 87 kN (permanent action).

Total = 366 kN (permanent action).

Vertical loads on crane leg (minimum)

Reaction from crane girder $V_{min} = 1357$ kN (permanent).

Weight of crane girder = 279 kN.

Self-weight of crane leg = 87 kN.

Total = 1723 kN.

The minimum loads on the roof leg are as follows.

DL (from roof + gas duct) = 181 + 480 = 661 kN.

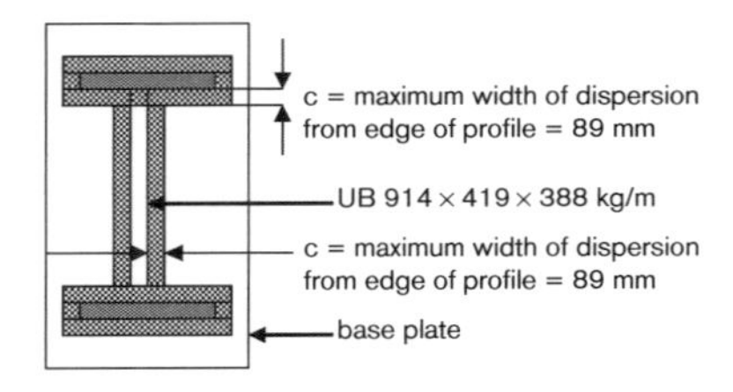

BASE PLAN

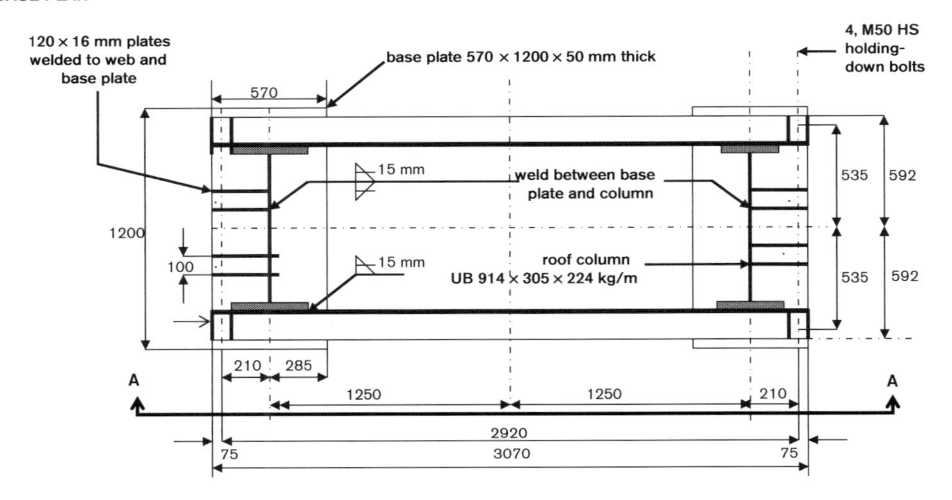

VIEW A–A

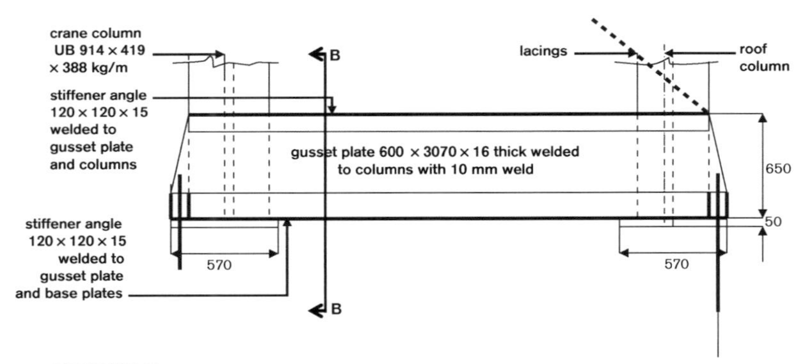

SECTION B–B

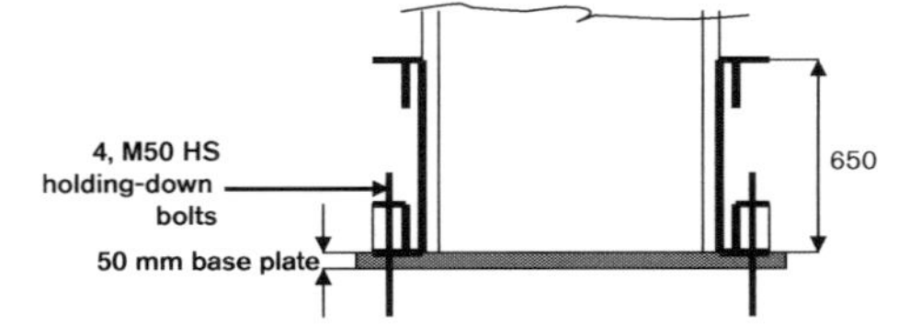

Notes:
(1) Grade of steel used is S275.
(2) For stanchion details along line A see Fig.8.3

Fig. 6.5. Details of base of stanchion

Self-weight of roof leg = 2.24 × 35.5 = 80 kN.

Total = 741 kN (permanent action).

Moments

When the wind and the crane surge are acting from right to left,

moment due to wind (from right to left) shared by the stanchion = 5310 kN m (anticlockwise);

moment due to crane surge (from right to left) = 3070 kN m (anticlockwise);

moment due to gas duct weight of 480 kN (2 m eccentricity) = −480 × 2 = −960 kNm (clockwise).

When the wind and the crane surge are acting from left to right,

moment due to wind (from left to right) shared by the stanchion = 4577 kN m (clockwise);

moment due to crane surge (from left to right) = 3070 kN m;

moment due to gas duct eccentricity = 960 kN m.

Ultimate thrust or tension in the legs

Case 1: when the wind and the crane surge are acting from right to left. We consider the thrust in the legs due to the moment with a 2.5 m spacing between the roof and crane legs.

Thrust due to wind = 5310/2.5 = ±2124 kN.

Thrust due to crane surge = 3070/2.5 = ±1228 kN.

Thrust due to gas duct = 960/2.5 = ±384 kN.

With a safety factor for permanent actions $\gamma_{Gj} = 1.35$ (DL), for the leading variable actions $\gamma_{wk} = 1.5$ (WL) and for the accompanying variable load $\gamma_{Cr} = 0.6 \times 1.5 = 0.9$ (crane surge),

ultimate resultant thrust in crane leg = 1.35 × 366 (DL) + 1.5 × 2124 (WL)
+ 0.9 × 1228 (crane surge) + 0.9
× 4062 (vertical crane girder reaction)
− 1.35 × 384 = 7923 kN.

Ultimate resultant tension in roof leg = 0.9 × (741 + 384) (DL + gas duct)
− 1.5 × 2124 (WL) − 0.9
× 1228 (crane surge) = −3279 kN (tension).

Case 2: when the wind and the crane surge are acting from left to right.

Thrust due to wind = ±4577/2.5 = ±1831 kN.

Thrust due to crane surge = ±3070/2.5 = ±1228 kN.

Thrust due to gas duct = 960/2.5 = ±384 kN.

Ultimate resultant tension in crane leg = 0.9 × 1723 (DL + min. crane girder reaction)
− 1.5 × 1831 (WL) − 0.9 × 1228 (crane surge)
− 0.9 × 384 (gas duct) = −2647 kN (tension).

Ultimate resultant compression in roof leg = 1.35 × 741 (DL) + 1.5 × 1831 (WL)
+ 0.9 × 1228 (crane surge) + 1.35
× 384 (gas duct) = 5370 kN.

From the above results, we find that *case 1* will give the maximum tension in the bolts.

6.3.5.3 Calculations for holding-down bolts

These calculations are based on Eurocode 3, Part 1-8 (Eurocode, 2005c). From the calculations above,

ultimate tension in roof leg = 3279 kN

(see Fig. 6.4). Take moments about the centre line of the crane leg. Let T be the ultimate tension in the anchor bolts. Therefore

$$T \times 2.71 = 3279 \times 2.5$$

$$T = 3279 \times 2.5/2.71 = 3025 \text{ kN}.$$

Referring to Table 3.1 ("Nominal values of the yield strength f_{yb} and the ultimate tensile strength f_{ub} for bolts") of Eurocode 3, Part 1-8 and Clause 3.3, for anchor bolts of steel grades conforming to Group 4 of the "1.2.4" reference standards, the ultimate tensile strength f_{ub} should not exceed 900 N/mm^2. If we use higher-grade bolts of class 10.9 with yield strength f_{yb} = 900 N/mm^2, then f_{ub} = 900 N/mm^2. Referring to Table 3.4 of Eurocode 3, Part 1-8 ("Design resistance for individual fasteners subjected to shear and/or tension"),

tension resistance of one anchor bolt = $F_{t,Rd} = k_2 f_{ub} A_s/\gamma_{M2}$

where $k_2 = 0.9$, A_s is the tensile stress area of the bolt or of the anchor bolt, f_{ub} is the ultimate tensile strength of the anchor bolt = 900 N/mm^2 and γ_{M2} is the partial safety factor = 1.25 (see the note in Table 2.1 of Eurocode 3, Part 1-8). Using M50 anchor bolts, the tension area at the shank is A_{net} = 1491 mm^2. Therefore

tension resistance/bolt = $0.9 \times 900 \times 1491/(1.25 \times 10^3) = 966$ kN

Using four M50 anchor bolts, $F_{t,Rd} = 966 \times 4 = 3864$ kN.

$F_{t,Ed}/F_{t,Rd} = 3025/3864 = 0.78 < 1$ Satisfactory

Therefore we provide four M50 HS bolts ($f_{yb} = f_{ub}$ = 900 N/mm^2).

6.3.6 Design of thickness and size of base plate (see Fig. 6.4)

We design the base plate by the effective-area method. When the size of the base plate is larger than that required to limit the nominal allowable bearing pressure of the concrete to $0.6f_{ck}$, where f_{ck} is the characteristic cylinder strength of the concrete, then a portion of its area should be taken as ineffective, according to Eurocode 2, Part 1-1, BS EN 1992-1-1: 2005 (Eurocode, 2005d). In our case, for a crane column of UB 914 × 419 × 388 kg/m, thickness of flange $t_f = 36.6$ mm; width of flange $b = 420.5$ mm;

thickness of web $t_w = 21.4$ mm; depth of section $h = 921.0$ mm.

We assume that the whole axial load is spread over an area of bedding material (in this case concrete) by dispersal through the base plate to a boundary at a distance c from the section profile. We assume also that the dispersion takes place over a distance of not less than the thickness of the flange $t_f = 36.6$ mm.

Firstly, we calculate the resultant compression on the crane leg (see Fig. 6.4, showing details of the stanchion base). We take moments about the centre of gravity of the group of anchor bolts on the right side. Let N_{rest} be the ultimate resultant compression in the crane leg. Then,

$$N_{rest} \times 2.71 = N_c \times 2.71 - N_t \times 0.5 = 7923 \times 2.71 - 3219 \times 0.21 = 20\,795$$

where N_c = ultimate resultant compression in the crane leg = 7923 kN

and N_t = ultimate resultant tension in tension leg = 3279 kN

(the values of N_c and N_t were calculated in Section 6.3.5.2). Therefore the compression on the base plate area is

$$N_{rest} = 20\,785/2.71 = 7674 \text{ kN (compression).}$$

Using grade 40C concrete,

ultimate cylinder strength of concrete $= f_{ck} = 40$ N/mm^2

bearing strength of concrete $= f_{bu} = 0.6 \times 40 = 24$ N/mm^2

bearing area required $= N_{rest}/f_{bu} = 7674 \times 10^3/24 = 319\,750$ mm^2

Let c be the maximum width of dispersion from the edge of the profile. So,

total length of dispersion along length of flange $= b + 2c = (420.5 + 2c)$ mm

width of dispersion across thickness of flange $= t_f + 2c = (36.6 + 2c)$ mm

length of dispersion along length of web $= h - 2c - 2t_f = (921 - 2 \times 36.6 - 2c)$ mm $= 847.8 - 2c$

width of dispersion across thickness of web $= t_w + 2c = (21.4 + 2c)$ mm

effective bearing area $= 2[(420.5 + 2c)(36.6 + 2c)] + (847.8 - 2c)(21.4 + 2c) = 319\,750$

or $c^2 + 670c - 67\,707 = 0$

$$c = [-670 \pm (670^2 + 4 \times 67\,707)^{0.5})]/2 = 89 \text{ mm.}$$

Therefore

thickness of base plate required $= t_p = c[3f_{bu}/f_y]^{0.5}$

where f_{bu} is the pressure under the base plate $= 0.6f_{cu} = 24$ N/mm^2, and f_y is the design strength of the base plate = 275 N/mm^2. For grade S 275 steel with a thickness greater than 80 mm, $f_y = 255$ N/mm^2. Therefore

t_p required = 89 × [3 × 24/275]$^{0.5}$ = 45.5 mm.

Therefore we adopt a thickness of the base plate t_p = 50 mm, a width B_p = 570 mm and a depth D_p = 1200 mm.

6.4 Design of stanchions along line B

6.4.1 Design considerations

Each stanchion is made up of two columns, one being the crane column and the other the roof column. They are spaced 3.0 m apart and laced together by means of lacings. The crane column carries the vertical and horizontal reactions from the gantry girder in the melting bay.

The roof column supports the roof trusses at 33.0 m level from the melting bay and at 43.5 m level from the storage hopper bay. The roof column also supports three hoppers containing 6930 kN of HBI (heavy briquette iron) material, stored to feed the melting furnace. The total load from the hoppers is shared equally by the roof columns of the storage hopper bay. In addition, the stanchions are subjected to wind loads. The total moment due to wind is shared by five stanchions of the structure (constituting the melting bay) in proportion to the stiffness of each stanchion (calculations shown previously). See Fig. 6.6 for the structural arrangement.

6.4.2 Loadings

6.4.2.1 Loadings on crane column (unfactored)

- Loading from maximum vertical crane girder reaction = V_{max} = 4062 kN.
- Loading from minimum vertical crane girder reaction = V_{min} = 1357 kN.
- Loading from crane girder self-weight = 279 kN.
- Loading from self-weight of crane column (assumed) = 103 kN.
- Loading from horizontal transverse crane girder surge = 122.8 kN (acting at 25.0 m level).

6.4.2.2 Loadings on roof column (unfactored)

- Loading from vertical DL reaction of roof truss directly on column at 33.0 m level = 44.39 kN.
- Loading from vertical LL reaction of roof truss directly on column at 33.0 m level = 121.5 kN.
- Loading from vertical DL reaction of roof girder at 33.0 m level = 45.6 × 3 = 136.8 kN.
- Loading from vertical LL reaction of roof girder at 33.0 m level = 121.5 × 3 = 364.5 kN.
- Loading from vertical DL reaction of roof truss directly on column at 43.5 m level = 19.73 kN.
- Loading from vertical LL reaction of roof truss directly on column at 43.5 m level = 54 kN.

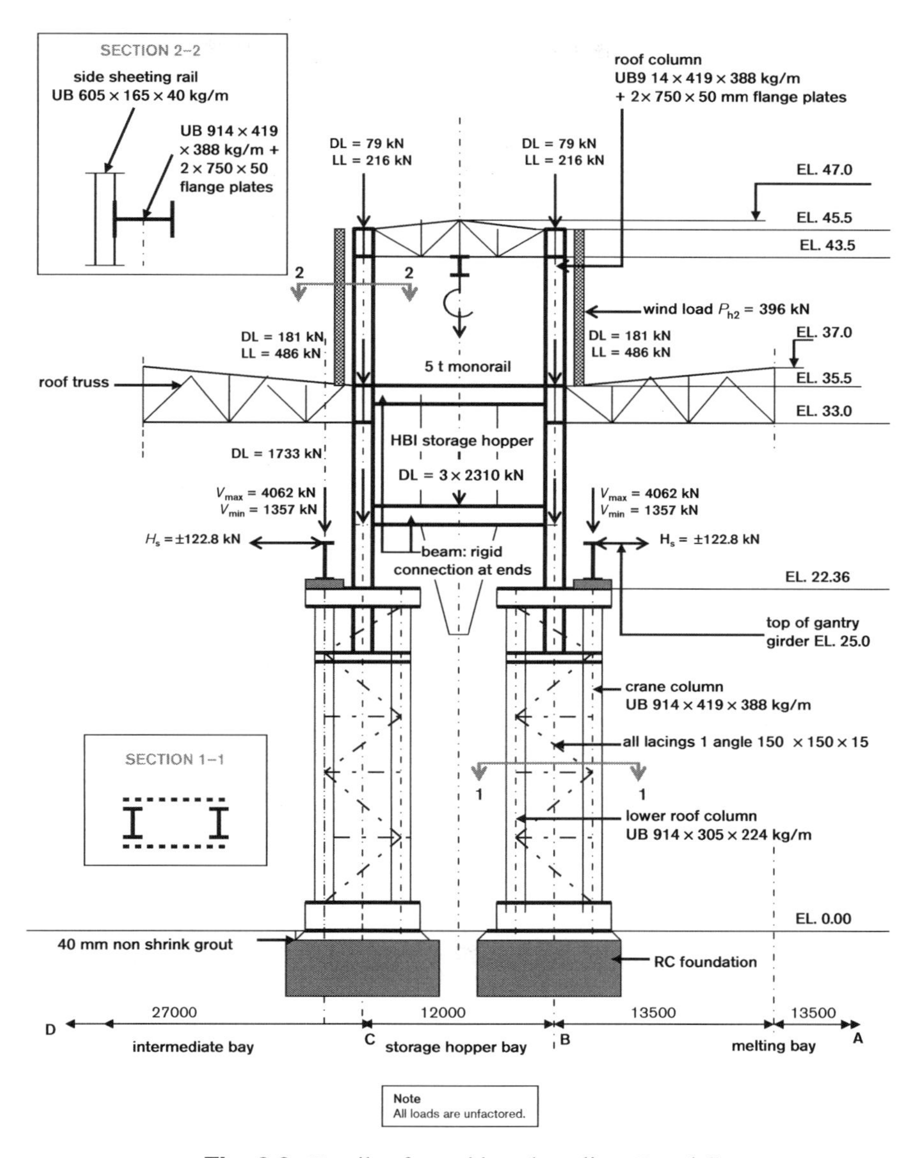

Fig. 6.6. Details of stanchion along lines B and C

- Loading from vertical DL reaction of roof girder at 43.5 m level = 19.73 × 3 = 59.2 kN.
- Loading from vertical LL reaction of roof girder at 43.5 m level = 54 × 3 = 162 kN.
- Loading from reaction of 3 hoppers (carrying 2310 kN each) = 3 × 2310/4 = 1732.5 kN.

Therefore

$$\sum \text{DL at 33.0 m level} = 44.39 + 136.8 = 181.19, \text{ say } 181 \text{ kN}.$$

$\sum$LL at 33.0 m level = 121.5 + 364.5 = 486 kN.

$\sum$DL at 43.5 m level = 19.73 + 59.2 = 78.93, say 79 kN.

$\sum$LL at 43.5 m level = 54 + 162 = 216 kN.

$\sum$DL on lower roof column = (181 + 79 + 1732.5)/2 + 2.24 × 25 (self-weight) = 1052 kN.

$\sum$LL on lower roof column = (486 + 216)/2 = 351 kN.

$\sum$DL on crane column = 1993/2 + 279 + 103 = 1378 kN.

$\sum$LL on crane column = 702/2 + 4062 = 4413 kN.

(A DL and an LL of 1993 kN and 702 kN, respectively, from the upper roof column will be shared equally by the crane and lower roof columns.)

6.4.3 Moments, unfactored

Case 1: when the wind and the crane horizontal surge are acting simultaneously from the right:

- moment shared by stanchion B due to wind (previously calculated) = 6288 kN m;
- moment due to crane horizontal surge at crane girder level (25.0 m) = 122.8 × 25 = 3070 kN m.

$\sum M$ = 6288 + 3070 = 9358 kN m.

Case 2: when the wind and the crane surge act simultaneously from the left:

- moment shared by stanchion B due to wind (calculated previously) = 5408 kN m;
- moment due to crane surge at crane girder level (25.0 m) = 3070 kN m.

$\sum M$ = 5408 + 3070 = 8478 kN m.

6.4.4 Thrust or tension due to unfactored moment from wind and crane surge

Case 1: when both wind and crane surge are acting from the right.

Spacing of crane column and lower roof column = 3.0 m.

Maximum thrust in roof column due to wind = 6288/3 = 2096 kN.

Maximum thrust in roof column due to surge = 3070/3 = 1023 kN.

Case 2: when both wind and crane surge act from left:

Maximum thrust in crane column due to wind = 5408/3 = 1803 kN.

Maximum thrust in crane column due to surge = 3070/3 = 1023 kN.

6.4.5 Ultimate design compression in crane and lower roof legs when DL + LL + WL and crane surge are acting simultaneously

For the partial safety factors, referring to BS EN 1990: 2002 (Eurocode, 2002), we assume $\gamma_{Gk} = 1.35$, $\gamma_{Qk} = 1.5$ (leading variable), $\gamma_{Wk} = 0.6 \times 1.5 = 0.9$ and $\gamma_{Cr,k} = 0.9$.

In *case 2*, the ultimate maximum design compression in the crane leg (when the wind is blowing from the left) is

$$N_{\mathrm{Ed}}\,(\text{crane leg}) = 1.35 \times \mathrm{DL} + 1.5 \times \mathrm{LL} + 0.9 \times \mathrm{WL} + 0.9 \times \text{crane load}$$

$$= 1.35 \times 1378 + 1.5 \times 4062 + 0.9 \times 1803 + 0.9 \times 1023 = 10\,497\ \mathrm{kN}.$$

In *case 2*, the ultimate maximum design compression in the lower roof leg (when the wind is blowing from the right) is

$$N_{\mathrm{Ed}}\,(\text{roof leg}) = 1.35 \times 1052 + 1.5 \times 351 + 0.9 \times 2096 + 0.9 \times 1023 = 4754\ \mathrm{kN}.$$

6.4.6 Design of section of columns in stanchion

6.4.6.1 Crane column

From the above calculations, ultimate maximum design compression = 10 497 kN.
Try a section UB914 × 419 × 388 kg/m: $A = 494\ \mathrm{cm}^2$, $i_y = 38.2$ cm, $i_z = 9.59$ cm.

Effective height along y-axis = $0.85L$ (top of column cap) = 0.85 × 22.36 = 19 m.

Effective height along z-axis = 3 m (spacing of lacings).

$\lambda_1 = 93.9\varepsilon = 93.9 \times 0.92 = 86.4$.

First, consider buckling about the y-axis:

$$\bar{\lambda} \text{ for } y\text{-axis} = L_{\mathrm{cry}}/(i_y\lambda_1) = 1900/(38.2 \times 86.4) = 0.58$$

Referring to Table 6.2 of Eurocode 3, Part 1-1, with $h/b = 921/420.5 = 2.19 > 1.2$ and $t_{\mathrm{f}} \leq 40$ mm, we follow the buckling curve “a” in Fig. 6.4 of that Eurocode. With $\bar{\lambda} = 0.58$, the reduction factor $\chi = 0.9$. Next, consider buckling about the z-axis:

$$\bar{\lambda} = L_{\mathrm{crz}}/(i_z\lambda_1) = 300/(9.59 \times 86.4) = 0.36$$

Referring again to Table 6.2 of Eurocode 3, Part 1-1, with $h/b > 1.2$, we follow the buckling curve “b” in Fig 6.4 of that Eurocode. With $\bar{\lambda} = 0.36$, $\chi = 0.95$. Thus, from the above we find that buckling about the y-axis will give a higher reduction factor. Therefore

$$N_{\mathrm{b,Rd}} = \chi A f_y/\gamma_{\mathrm{M1}} = 0.9 \times 494 \times 100 \times 275/103 = 12\,227\ \mathrm{kN} > N_{\mathrm{Ed}}\ (10\,497\ \mathrm{kN})$$

$$N_{\mathrm{Ed}}/N_{\mathrm{b,Rd}} = 10\,497/12\,227 = 0.86 < 1.0. \text{ So, OK.}$$

Therefore we adopt UB914 × 419 × 388 kg/m for the column supporting the crane.

6.4.6.2 Lower roof column

Ultimate maximum design compression = 4754 kN.
Try UB914 × 305 × 224 kg/m: $A = 286\ \mathrm{cm}^2$, $i_y = 36.3$ cm, $i_z = 6.27$ cm.

$L_{\mathrm{cry}} = 0.85 \times 22.36 = 19$ m, $L_{\mathrm{crz}} = 3$ m,

$\lambda_1 = 86.4$.

First, consider buckling about the y-axis:

$$\bar{\lambda} = L_{\mathrm{cry}}/(i_y\lambda_1) = 1900/(36.3 \times 86.4) = 0.61$$

Referring to Table 6.2 of Eurocode 3, Part 1-1, with $h/b > 1.2$ and $t_f \leq 40$ mm, we follow the buckling curve "a" in Fig. 6.4 of that Eurocode. We find that the reduction factor $\chi = 0.88$. Next, consider buckling about the z-axis:

$$\bar{\lambda} = 300/(6.27 \times 86.4) = 0.55$$

Referring again to Table 6.2 of Eurocode 3, Part 1-1, with $h/b > 1.2$, we follow the buckling curve "b" in Fig. 6.4 of that Eurocode. We find that the reduction factor $\chi = 0.84$. Thus, the z-axis will give a lower value of the compression resistance. Therefore

$$N_{b,Rd} = \chi A f_y/\gamma_{M1} = 0.84 \times 286 \times 275/10 = 6607 \text{ kN} > N_{Ed}$$

$$N_{Ed}/N_{b,Rd} = 4754/6607 = 0.72 < 1.0 \qquad \underline{\text{OK}}$$

Therefore we adopt UB914 × 305 × 224 kg/m for the lower roof column leg (see Fig. 6.6).

6.4.6.3 Upper roof column

We assume that the column is fixed at 22.36 m level and acts as a portal frame between the 22.5 and 35.5 m levels. The column is cantilevered between the 35.5 and 43.5 m levels. The moment shared by the stanchion B due to wind blowing from the right at the base (level 0.00) is $M_b = 6288$ kN m. Assuming that the moment diagram is of triangular form, the moment at 22.36 m level from the base is

M_b (22.36) = 6288 × 22.36/45.5 = 3090 kN m.

Vertical DL = 79.0 + 181 + 1732.5 + 3.88 × (45.5 − 22.36) (self-weight) = 2083 kN.

Vertical LL = 486 + 216 = 702 kN (already calculated).

With partial safety factors $\gamma_{Gk} = 1.35$, $\gamma_{Qk} = 15 \times 0.6 = 0.9$ (accompanying variable) and $\gamma_{wk} = 1.5$ (leading variable),

maximum ultimate design compression $N_{Ed} = 1.35 \times 2083 + 0.9 \times 702 = 3443$ kN;

maximum ultimate design moment $M_{Ed} = 1.5 \times 3090 = 4635$ kN m.

Try UB914 × 419 × 388 kg/m + two flange plates 750 × 50 mm: $A = 1254$ cm^2, $i_y = 44.7$ cm, $i_z = 17.9$ cm, $W_y = 48\,710$ cm^3.

To check for direct compression

We assume that the member is rigidly connected to the beams at 28.5 and 35.5 m levels. So, the free-standing height will be from 35.5 to 43.5 m. Therefore $L_{cry} = 1.5 \times 8 = 12$ m and $L_{crz} = 8$ m (as the inside compression flange is restrained between 35.5 and 43.5 m levels). Therefore

$$\lambda_1 = 86.4,$$

$$\bar{\lambda}_y = L_{cry}/(i_y \lambda_1) = 1200/(44.7 \times 86.4) = 0.32,$$

$$\bar{\lambda}_z = L_{crz}/(i_y \lambda_1) = 800/(17.9 \times 86.4) = 0.50.$$

Referring to Table 6.2 of Eurocode 3, Part 1-1, with $t_f > 40$ mm, we follow the curve "c" for buckling about the y-axis and the curve "d" for buckling about the z-axis in Fig. 6.4 of that Eurocode, and find that $\chi_y = 0.95$ and $\chi_z = 0.8$. Therefore

$N_{b,Rd} = \chi_z A f_y / M_1 = 0.8 \times 1244 \times 275/10 = 27368$ kN > 3443 kN

$N_{Ed}/N_{b,Rd} = 3443/27368 = 0.13 < 1.0$ OK

To check the shear capacity

Total shear at 35.5 m level = 396 kN.

Ultimate shear = $V_{Ed} = 1.5 \times 396 = 594$ kN.

Assume that the shear is shared equally by roof columns B and C. Therefore

ultimate shear in column B = 594/2 = 297 kN

Shear resistance = $V_{b,Rd} = A_v(f_y/\sqrt{3})/\gamma_{Mo}$

$A_v = A - 2b_{tf} + (t_w + 2r)t_f = 49\,400 - 2 \times 420.5 \times 36.6 + (21.4 + 2 \times 24.1) \times 36.6$
$= 21\,167$ mm^2

Therefore

$V_{b,Rd} = 21\,167 \times (275/(3^{0.5})/10^3 = 3361$ kN > 594 kN OK

To check the moment capacity

Where the shear is less than half the plastic shear resistance, the moment capacity need not be reduced. In our case the ultimate shear, 594 kN, is less than half the plastic shear resistance, 3361/2 = 1680 kN. So, the moment resistance is

$My_{,Rd} = W_{pl} f_y / \gamma_{Mo} = 48\,710 \times 275/10^3 = 13\,395$ kN m > 4635 kN m OK

To check for lateral–torsional buckling

The effective height in the y–y direction is $L_{cry} = 19.5$ m. Referring to Clause 6.3.2.4 of Eurocode 3, Part 1-1

$$\bar{\lambda}_f = (k_c L_c)/(i_{f,z}\lambda_1) \leq \bar{\lambda}_{c0}(M_{c,Rd}/M_{y,Ed}) \qquad (6.59)$$

where $My_{,Ed}$ is the maximum design value of the bending moment within the restraint spacing, equal to 4635 kN m;

$M_{c,Rd} = W_y f_y / \gamma_{M1} = 48\,710 \times 275/10^3 = 13\,395$ kN m

W_y is the section modulus of the section about the y–y axis, equal to 48 710 cm^3; k_c is the slenderness correction factor (from Table 6.6 of Eurocode 3, Part 1-1, $k_c = 1.0$); L_c is the unrestrained height of the compression flange for lateral–torsional buckling, equal to 1200 cm; $i_{f,z}$ is the radius of gyration of the equivalent compression flange, equal to 17.9 cm; and $\bar{\lambda}_{c0}$ is the slenderness limit of the equivalent compression flange. Referring to Clause 6.3.2.3,

$\bar{\lambda}_{c0} = \bar{\lambda}_{LT,0} + 0.1 = 0.4$ (recommended) $+ 0.1 = 0.5$

$\lambda_1 = 93.9\varepsilon = 93.9 \times 0.92 = 86.4$

Therefore

$$\bar{\lambda}_f = (k_c L_c)/(i_{f,z}\lambda_1) = 1.0 \times 1200/(17.9 \times 86.4) = 0.78$$

$$\text{and} \quad \bar{\lambda}_{c0}(M_{c,Rd}/My_{,Ed}) = 0.5 \times 13\,395/4635 = 1.44 > 0.78$$

So, no reduction of buckling resistance moment needs to be considered.

$$N_{Ed}/N_{b,Rd} = 3443/27\,368 = 0.13$$

Therefore

$$N_{Ed}/N_{b,Rd} + M_{Ed}/M_{c,Rd} = 0.13 + 4635/13395 = 0.13 + 0.35 = 0.48 < 1$$

Therefore we adopt UB914 × 419 × 388 kg/m + two flange plates 750 × 50 mm for the upper portion of the stanchion.

References

Eurocode, 2002. BS EN 1990: 2002(E), Basis of structural design.

Eurocode, 2005a. BS EN 1991-1-4: 2005, Actions on structures. General actions. Wind actions.

Eurocode, 2005b. BS EN 1993-1-1: 2005, Eurocode 3. Design of steel structures. General rules and rules for buildings.

Eurocode, 2005c. BS EN 1993-1-8: 2005, Eurocode 3. General. Design of joints.

Eurocode, 2005d. BS EN 1992-1-1: 2005, Eurocode 2. General. Common rules for building and civil engineering structures.

CHAPTER 7

Case study II: Design of Gable End Framing System Along Row 10, Based on Eurocode 3

7.1 Design considerations (see Figs 1.1 and 7.1)

The gable end framing system consists of a series of vertical gable columns with a regular spacing, supported at the bottom on a ground level foundation and at the top on the bottom chord level of the roof truss. The columns are continuous over the horizontal wind girder provided at about the cap level (22.5 m) of the stanchions. The top and bottom of the gable columns are assumed to be hinged. The gable columns are spaced at 4.5 m intervals in the melting bay, and at 6.0 m intervals in the storage hopper bay above 22.5 m level.

The horizontal wind girder at 22.5 m level spans between the crane columns and is a lattice girder. It is kept horizontal by brackets from the vertical gable columns as shown in Fig. 1.1. The depth of the girder is assumed to be 1/10 of the span. So, the depth is 2.3 m.

The horizontal wind girder at the bottom chord level of the roof truss at 33.0 m level is formed by latticing between the trusses of rows 9 and 10 as shown in Fig. 7.1. Thus the depth of the truss is 6.0 m.

A door and windows have been provided in the shop for the movement of traffic and ventilation, respectively.

7.2 Functions

The functions of the gable columns are to support the side rails, which are subjected to wind loads and the weight of the side covering. The wind girders support the reactions from the gable columns. So, we have to design the gable columns and wind girders for the conditions mentioned above.

7.3 Design of gable columns

7.3.1 Design data

Spacing = 4.5 m.

Height = 33.0 m, spanning over the intermediate wind girder at 22.52 m level.

7.3.2 Loadings

The gable columns are subjected to wind loads in addition to the weight of the side covering.

- Dead loads:

GABLE FRAMING

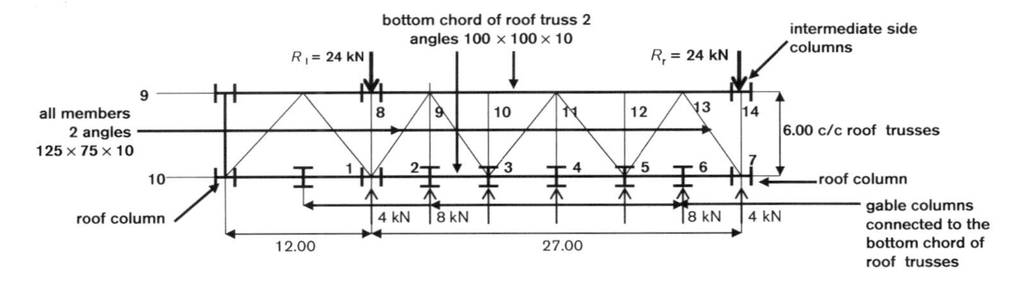

PLAN OF WIND GIRDER AT BOTTOM CHORD OF ROOF TRUSS

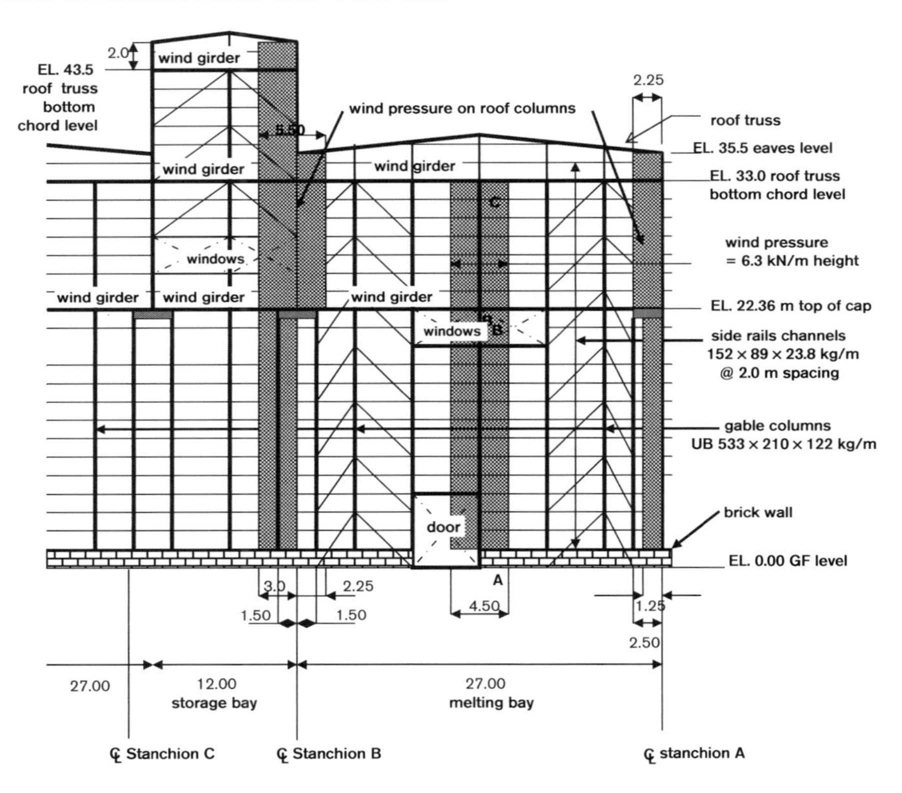

PLAN OF WIND GIRDER AT 22.36 m LEVEL

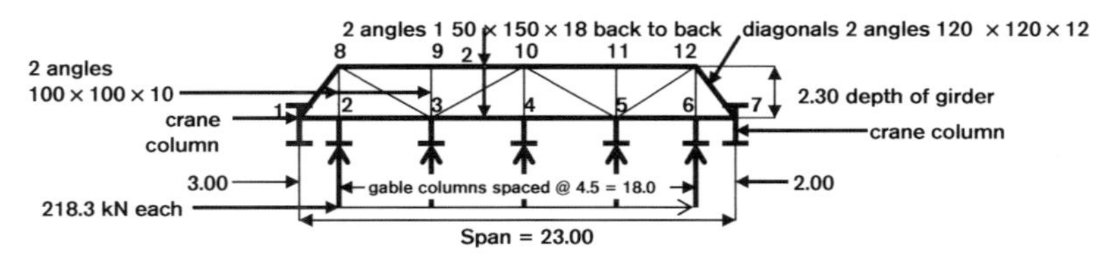

Fig. 7.1. Gable end framing system along row 10

as calculated before, weight of sheeting = 0.258 kN/m^2;

dead load/m height of column (spacing = 4.5 m) = 0.258 × 4.5 = 1.16 kN/m height;

self-weight of column (assume UB 533 × 210 × 92 kg/m) = 0.92 kN/m.

Total = 2.08 kN/m.

Therefore

total vertical load on column = G_k = 2.08 × 33 = 68.67 kN.

- Wind loads:

External wind pressure on wall = $p_e = q_p c_{pe}$ = 1.27 × 1.1 = 1.4 kN/m^2

(previously calculated, see Chapter 2).
Therefore

external wind pressure/m height of column = w_k = 1.4 × 4.5 = 6.3 kN/m

(as the columns are spaced 4.5 m apart).

7.3.3 Moments

As the member is continuous over the intermediate support, we consider it as an indeterminate member. To obtain the moment at the support and at midspan, we carry out a moment distribution analysis, as follows.

7.3.3.1 Step 1: To calculate the stiffness $K = I/L$ of the member

Let ABC be the member, with the ends A and C simply supported, and continuous over the support at B. The lengths of the parts are L_{AB} = 22.36 m and L_{BC} = 10.64 m. Let I be the moment of inertia of the member. Then, the stiffnesses of the parts of the member are $K_{AB} = I/L_{AB}$ and $K_{BC} = I/L_{BC}$. Since the ends A and C are simply supported, the stiffnesses are taken as three-quarters of these values. Therefore

$K_{AB} = (3/4)I/L_{AB} = 0.75I/22.36 = 0.033$,

$K_{BC} = (3/4)I/L_{BC} = 0.75I/10.64 = 0.071$.

7.3.3.2 Step 2: To calculate the distribution factors DF of the parts of the member meeting at the joint

Consider the intermediate support point B where the parts AB and BC meet. The distribution factor for BA at the support B is $DF_{BA} = K_{AB}/(K_{AB} + K_{BC})$. Therefore

$DF_{BA} = 0.033/(0.033 + 0.071) = 0.033/0.105 = 0.31$

and the distribution factor for BC at the support B is

$DF_{BC} = 0.071/(0.033 + 0.071) = 0.69$.

7.3.3.3 Step 3: To calculate the fixed-end moments at the ends of the parts of the member, assuming no rotation at the ends takes place

The fixed-end moments at the ends of AB are equal to $w_k L_{AB}^2/12$. Since the support A is simply supported, 50% of the moment is transferred to B and the moment at A is reduced to zero. Thus, consider the support point B. For AB, the fixed-end moment at B is

$$1.5 w_k L_{AB}^2/12 = w_k L_{AB}^2/8 = 6.3 \times 22.36^2/8 = 394 \text{ kN m}$$

and the fixed-end moment at A is zero. Similarly, for BC, the fixed-end moment at B is

$$w_k L_{BC}^2/8 = 6.3 \times 10.64^2/8 = 89 \text{ kN m}$$

and the fixed-end moment at C is zero.

7.3.3.4 Step 4: To find the moments

We carry out a moment distribution analysis as shown in Table 7.1.

7.3.3.5 Step 5: Field moment and shear at the supports

From the results in Table 7.1, we calculate the field moment and shear at the supports:

$$\text{net positive BM at midspan of AB} = w_k L_{AB}^2/8 - M_{BA}/2 = 6.3 \times 22.36^2/8 - 299.5/2 = 244 \text{ kN m};$$

$$\text{net positive BM at midspan of BC} = w_k L_{BC}^2/8 - M_{BC}/2 = 6.3 \times 10.64^2/8 - 299.5/2 = -60.6 \text{ kN m};$$

net BM at 2.5 m above intermediate support B
$= 6.3 \times 2.5/2 \times (1 - 2.5/10.64) - 299.5 \times 3/4 = 67 - 224.6 = 157.6$ kN m.

Reaction at support B
$= R_b = w_k(L_{AB} + L_{BC})/2 + M_{BA}/22.36 = 6.3 \times 33/2 + 299.5/22.36 + 299.5/10.64$
$= 145.5$ kN.

Reaction at support C
$= R_c = w_k L_{BC}/2 - M_{BC}/10.64 = 6.3 \times 10.64/2 - 299.5/10.64 = 5.4$ kN.

7.3.3.6 Step 6: Ultimate design moments, shear and thrust

Thus, the member is subjected to a vertical thrust due to dead load $N = 68.67$ kN, a maximum moment due to wind $M_{BA} = 299.5$ kN m and a shear at B due to wind

$$V = w_k L_{AB}/2 + M_{BA}/22.36 = 83.8 \text{ kN}.$$

Table 7.1. Analysis of bending moment by the Hardy Cross moment distribution method (Butterworth, 1949).

	Support A	Support B		Support C
Distribution factor		0.31	0.69	
Fixed-end moment		+394	−89	0
Distribution		−94.5	−210.5	
Final moment (kN m)	0.0	+299.5	−299.5	0.0

Now, we have to consider the ultimate design values with a partial safety factor. When we consider that the wind load acts simultaneously with the dead load, the partial safety factor γ_{wk} should be taken equal to 1.5. But, from the above results, we find that the wind load is dominant compared with the dead load. So, in this case we adopt a partial safety factor for the wind load $\gamma_{wk} = 1.5$ and a partial safety factor for the dead load $\gamma_{Gk} = 1.35$.

Ultimate maximum design value for vertical load $N_{Ed} = 1.35 \times 68.67 = 93$ kN.

Ultimate maximum design value for moment $M_u = M_{Ed} = 1.5 \times 299.5 = 449$ kN m

at intermediate support.

Ultimate maximum design value for moment at midspan of BC $= -1.5 \times 60.6 = 91$ kN m.

Assume that the average moment at a point 2.5 m above the intermediate support B is $157.6 \times 1.5 = 236$ kN m.

Ultimate maximum design value for shear $V_{Ed} = 1.5 \times 83.8 = 126$ kN.

Fig. 7.2 shows the deflected shape and the BM and SF diagrams.

7.3.4 Design of section, based on Eurocode 3, Part 1-1 (Eurocode, 2005)

(All equation numbers in this subsection herein refer to Eurocode 3 unless otherwise mentioned.) The deflected shape of the member and the BM diagram in Fig. 7.2 show that the outer compression flange connected to the side rails is restrained from lateral buckling up to a height of about 18.36 m from the base, whereas for the rest of the height the

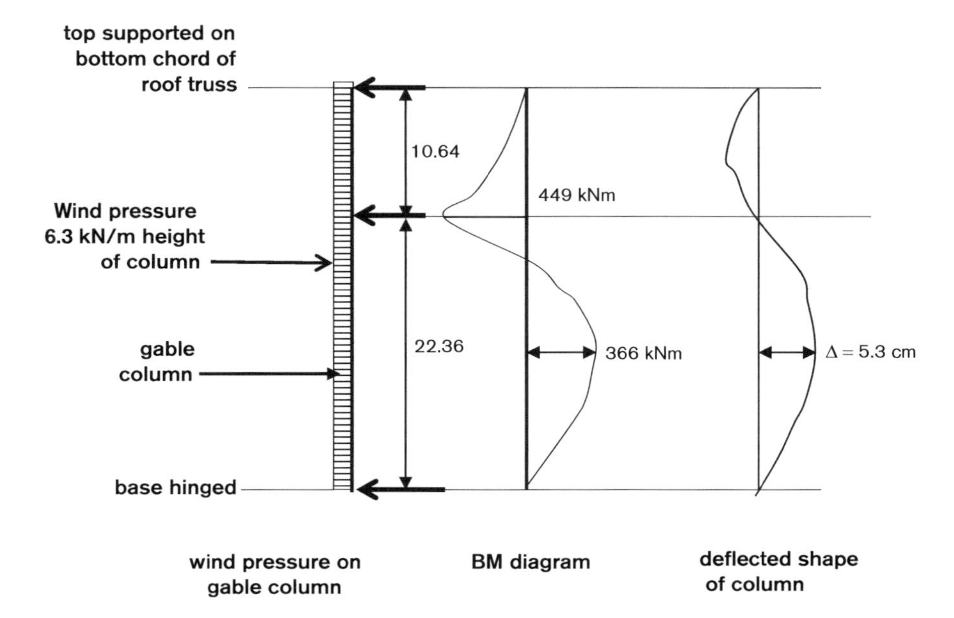

Fig. 7.2. Gable column: BM diagram and deflected shape

inner flange is subjected to compression. The compression flange is unrestrained in this range height of the member, and torsional buckling of the member can occur, thus reducing the buckling resistance of the member. The unrestrained height of the inner portion of the compression flange of the member AB may be assumed to be at the point of contraflexure, as can be seen from the bending moment diagram.

For the upper portion BC of the column, 10.64 m in height, the inner flange is also under compression. As this portion of the inner flange is unrestrained, a torsional buckling stress will be developed.

With the above design concept and with the values of moment and thrust calculated above, the following sequence of steps was followed in the design of the member.

7.3.4.1 Step 1: Selection of section, and properties of the section

Try a section UB 533 × 210 × 82 kg/m: $t_f = 13.2$ mm, $t_w = 9.6$ mm, $h = 528.3$ mm, $b = 208.8$ mm, $A_g = 105$ cm^2, $W_{yy} = 2060$ cm^3, $W_{zz} = 300$ cm^3, $i_y = 21.3$ cm, $i_z = 4.38$ cm, $h_w = 476.5$ mm. Assume steel grade S 275, yield strength $p_y = 275$ N/mm^2.

7.3.4.2 Step 2: Classification of cross-section

Before designing the section, we have to classify the section into one of the following classes:

- *Class 1, plastic:* cross-section with plastic hinge rotation capacity.
- *Class 2, compact:* cross-section with plastic moment capacity.
- *Class 3, semi-compact:* cross-section in which the stress in the extreme compression fibre can reach the design strength, but a plastic moment capacity cannot be developed.
- *Class 4, slender:* cross-section in which we have to have a special allowance owing to the effects of local buckling.

Flange

Stress factor $\varepsilon = (235/f_y)^{0.5} = (235/275)^{0.5} = 0.92$.

Outstand of flange $c = (b - t_w - 2r)/2 = (208.8 - 9.6 - 2 \times 12.7)/2 = 86.9$ mm.

Ratio $c/t_f = 86.9/13.2 = 6.6$.

For class 1 classification, the limiting value of $c/t_f \leq 9\varepsilon$. We have $6.6 \leq 9 \times 0.92$, i.e. $6.6 \leq 8.28$. So, the flange satisfies the conditions for class 1 classification. OK

Web

Ratio $h_w/t_w = 476.5/9.6 = 49.6$

For class 1 classification, $h_w/t_w \leq 72\varepsilon$. We have $49.6 \leq 72 \times 0.92$, i.e. $49.6 \leq 66.2$. So, the web satisfies the conditions for class 1 section classification.

7.3.4.3 Step 3: To check the shear capacity of the section

The ultimate design shear force V_{Ed} should not be greater than the design plastic shear resistance $V_{pl,Rd}$ of the section. The plastic shear resistance is given by the following equation:

$$V_{pl,Rd} = A_v(f_y/\sqrt{3})/\gamma_{Mo} \quad (6.18)$$

where A_v = shear area of section = $A - 2bt_f + (t_w + 2r)t_f$

$= 105 \times 100 - 2 \times 208.8 \times 13.2 + (9.6 + 2 \times 12.7) \times 13.2 = 5450\ mm^2$.

$V_{pl,Rd} = 5450 \times (275/\sqrt{3})/1/10^3 = 865\ kN > V_{Ed}\ (231\ kN)$. Satisfactory

7.3.4.4 Step 4: To check for shear buckling

Referring to Table 5.2 of Eurocode 3, Part 1-1, if the ratio d_w/t_w exceeds 72ε, then the web should be checked for shear buckling. In our case,

$h_w/t_w = 49.6$

and $72\varepsilon = 66.2 > h_w/t_w$.

Therefore the web need not be checked for shear buckling.

7.3.4.5 Step 5: To check the moment capacity of the section

Referring to Section 6.2.8 of Eurocode 3, Part 1-1, where a shear force and bending moment act together, the presence of the shear force has an effect on the moment resistance of the member. If the shear force is less than half the plastic shear resistance, its effect on the moment resistance may be neglected except when shear buckling reduces the section resistance. In our case, the shear force $V_{ed} = 126$ kN and the plastic shear resistance $V_{pl,rd} = 865$ kN (as calculated before). From these values, we see that the shear force V_{Ed} is less than half of the shear resistance $V_{pl,rd}$. Thus, the effect of the shear force on the moment resistance can be neglected. So, the design value of the bending moment M_{Ed} at each cross-section should satisfy

$$M_{Ed}/M_{c,Rd} \leq 1.0 \quad (6.12)$$

where

$$M_{c,Rd} = M_{pl,Rd} = \text{design resistance for bending} = W_{pl}f_y/\gamma_{Mo} \quad (6.13)$$

and where W_{pl} is the plastic section modulus. In our case, $W_{pl} = 2060\ cm^3$. Therefore

$M_{pl,Rd} = 2060 \times 10^3 \times 275/10^6 = 567\ kN\ m$;

$M_{Ed}/M_{pl,Rd} = 449/567 = 0.79 < 1$. OK

7.3.4.6 Step 6: To check the lateral–torsional buckling resistance moment

The gable end columns sag at midspan and hog near the intermediate support at 22.36 m level. We assume that the compression flange on the outer face at midspan is restrained by the fixings of the side rails. But near the intermediate support, the compression flange is on the inside face of the building. So, to restrain the inside compression flange, a strut member is introduced at about 5.64 m below the top wind girder level (33.0 m level). We have also assumed that the point of contraflexure at about 4 m below the intermediate wind girder level (22.36 m) is a point of restraint, at the unrestrained height of the member. So, we calculate based on the unrestrained height of 5 m mentioned above.

The average design moment at 2.5 m above the intermediate support is $M_{Ed} = 236$ kN m (calculated above). So, the effective height of the column in the z–z direction is 500 cm and the effective height in the y–y direction is 2250 cm.

Referring to Clause 6.3.2.4 of Eurocode 3, Part 1-1, members with discrete lateral restraint to the compression flange are not susceptible to lateral–torsional buckling if the length L_c between restraints or the resulting slenderness $\bar{\lambda}_f$ of the equivalent compression flange satisfies the following equation:

$$\bar{\lambda}_f = k_c L_c / (i_{f,z} \lambda_1) \leq \bar{\lambda}_{c,0} (M_{c,Rd} / M_{y,Ed}), \qquad (6.59)$$

where $M_{y,Ed}$ is the maximum design bending moment within the restraint spacing, equal to 236 kN m;

$$M_{c,Rd} = W_y f_y / \gamma_{M1} = 2060 \times 275/10^3 = 567 \text{ kN m};$$

L_c is the length between restraints, equal to 500 cm; and k_c is the slenderness correction factor for the moment distribution within the restraint spacing, equal to 0.91. Referring to Table 6.6 ("Correction factor") of Eurocode 3, Part 1-1,

$i_{f,z}$ = radius of gyration of compression flange composed of compression flange + 1/3 of compression part of web = 4.38 cm (web neglected);

$$\bar{\lambda}_{c,0} = \bar{\lambda}_{LT,0} + 0.1 = 0.4 + 0.1 = 0.5$$

(see Clause 6.3.2.3 of Eurocode 3, Part 1-1), and

$$\lambda_1 = 93.9\varepsilon = 6.4.$$

Therefore

$$\bar{\lambda}_f = k_c L_c / (i_{f,z} \lambda_1) = 0.91 \times 500/(4.38 \times 86.4) = 1.2,$$

$$\bar{\lambda}_{c,0} (M_{c,Rd} / M_{y,Ed}) = 0.5 \times 567/236 = 1.2.$$

Thus, no reduction of buckling resistance moment will be considered. Therefore

buckling resistance moment $M_{b,Rd} = M_{c,Rd} = 567$ kN m.

7.3.4.7 Step 7: To check for buckling resistance as a compression member

In addition, the member is also subjected to compression, and the effect of the combined buckling resistance to compression and buckling moment resistance should not exceed the allowable stress in the member. The gable end column is assumed to be hinged at the bottom, and continuous over the intermediate horizontal lattice girder at 22.36 m level. We shall check the design buckling resistance of this column as a compression member. Referring to the equations

$$N_{b,Rd} = \chi A f_y / \gamma_{M1} \qquad (6.47)$$

$$\text{and} \quad \bar{\lambda} = L_{cr} / (i_y \lambda_1), \qquad (6.50)$$

(see Clauses 6.3.1.1 and 6.3.1.3 of Eurocode 3, Part 1-1), we obtain

$$\lambda_1 = 93.9\varepsilon = 86.4,$$

L_{cr} = effective height of column from ground floor to intermediate support on y–y axis = 0.85 × 2236 = 1901 cm (as the member is assumed hinged at the base and continuous over the intermediate support, the effective height is taken equal to $L_{cr} = 0.85L$), and

$$\bar{\lambda} = 1901/(21.3 \times 86.4) = 1.03.$$

Referring to Table 6.2 of Eurocode 3, Part 1-1, with $h/b > 1.2$ and $t_f \leq 40$, we follow the curve "a" in Fig. 6.4 of that Eurocode. With $\bar{\lambda} = 1.03$, the reduction factor $\chi = 0.64$. Therefore

$$N_{b,Rd} = \chi A f_y / \gamma_{M1} = 0.64 \times 105 \times 275/10 = 1848 \text{ kN}.$$

Therefore

$$My_{,Ed}/M_{b,Rd} + N_{Ed}/N_{Rd} = 236/567 + 93/1848 = 0.41 + 0.05 = 0.46 < 1.$$ Satisfactory

Therefore we adopt UB533 × 210 × 82 kg/m.

7.3.4.8 Step 8: To check deflection

The section must be checked for deflection in the serviceability limit state. So, the wind load per metre height of column is 6.3 kN/m. The member is continuous over the intermediate wind girder at a height of 22.36 m from ground level. Consider the member to be hinged at the base and fixed at 22.5 m level. Therefore

$$\Delta_{act} = w_k l^4/(185 E I_y) = 6.3 \times 22.36^4 \times 100^3/(185 \times 21\,000 \times 47\,500) = 8.5 \text{ cm},$$

where $w_k = 6.3$ kN/m, $E = 21\,000$ kN/cm^2 and $I_y = 47\,500$ cm^4. The allowable deflection is

$$\Delta_{allowable} = l/360 = 22.36 \times 100/360 = 6.2 \text{ cm} < \Delta_{act}.$$

Therefore we increase the section from UB533 × 210 × 82 kg/m to UB533 × 210 × 122 kg/m (see Fig. 7.1).

7.4 Design of horizontal wind girder at 22.36 m level

7.4.1 Design considerations (see Fig. 7.1)

The horizontal wind girder is of lattice type construction, and spans 23 m between the crane columns of stanchions A and B. It supports the gable columns at 22.36 m level and carries the reactions from the gable columns.

Depth of girder assumed = 2.3 m.

7.4.2 Loadings

Reactions from gable column at node point = 145.5 kN (unfactored).
Assuming a partial safety factor $\gamma_{wk} = 1.5$ (for wind alone),

ultimate node point load = 1.5 × 145.5 = 218.3 kN.

7.4.3 Forces in lattice members of girder

Consider the lattice girder (1-7-8-12), as shown in Fig. 7.1, with loads of 218.3 kN at node points 2, 3, 4, 5 and 6.

Reaction at right support $R_7 = 218.3 \times (3 + 7.5 + 12 + 16.5 + 21)/23 = 569.5$ kN.

Reaction at left support $R_1 = 5 \times 218.3 - 569.5 = 522$ kN.

Consider node 7 (right).

$\sum V = 0$:

(7-12) × 2.3/3.05 = 569.5; force in member (7-12) = 569.5 × 3.05/2.3 = 755 kN (tension).

$\sum H = 0$:

(6-7) = (7-12) × 2/3.05; force in member (6-7) = 755 × 2/3.05 = 495 kN (compression).

Next, consider node 12.

$\sum V = 0$:

(7-12) × 2.3/3.05 − (6-12) − (5-12) × 2.3/5.05 = 0;

force in member (5-12) = 771 kN (compression).

$\sum H = 0$:

(7-12) × 2/3.05 + (5-12) × 4.5/5.05 − (11-12) = 0; force in member (11-12) = 1182 kN (tension).

(Force in member (6-12) = 218.3 kN (compression).)

Next, consider node 5.

$\sum V = 0$:

218.3 − (5-12) × 2.3/5.05 + (5-10) × 2.3/5.05 = 0; force in member (5-10) = 292 kN (tension).

$\sum H = 0$:

−(5-6) + (4-5) − (5-12) × 4.5/5.05 − (5-10) × 4.5/5.05 = 0;

force in member (4-5) = 1441 kN (compression).

(Force in member (5-11) = 0.)

See Table 7.2 for the forces in the members and the sizes of those members.

7.4.4 Design of sections

First, we consider the diagonal member (5-12).

Force $N_{Ed} = 771$ kN (compression).

To design this member as a compression member, we try a section made from two angles 120 × 120 × 12 back to back with a 12 mm gap between the vertical faces: $A = 55$ cm^2, $r_y = 2.72$ cm,

L_{cr} = effective length = $L = 305$ cm.

Table 7.2. Forces in and sizes of members

Member	Location	Force (kN), ultimate[a]	Member size, 2 angles $h \times b \times t$[b]
6-7, 1-2	Outer chords	495 (C)	2 × 150 × 150 × 18
5-6, 2-3	Outer chords	495 (C)	2 × 150 × 150 × 18
4-5, 3-4	Outer chords	1441 (C)	2 × 150 × 150 × 18
11-12, 8-9	Inner chords	1182 (T)	2 × 150 × 150 × 18
10-11, 9-10	Inner chords	1182 (T)	2 × 150 × 150 × 18
7-12, 1-8	Diagonals	755 (T)	2 × 120 × 120 × 12
5-12, 3-8	Diagonals	771 (C)	2 × 120 × 120 × 12
5-10, 3-10	Diagonals	292 (T)	2 × 120 × 120 × 12
6-12, 2-8	Horizontals	218.3 (C)	2 × 100 × 100 × 10
5-11, 3-9	Horizontals	0	2 × 100 × 100 × 10
4-10	Horizontal	218.3 (C)	2 × 100 × 100 × 10

[a] Compression forces are denoted by (C) and tension forces by (T).
[b] b = horizontal dimension of angle; h = vertical dimension of angle; t = thickness of angle.

For the section classification,

$$f_y = 275 \text{ N/mm}^2,\ \varepsilon = (235/f_y)^{0.5} = 0.92,\ h/t = 120/12 = 10.$$

$$15\varepsilon = 15 \times 0.92 = 13.8,\ 11.5\varepsilon = 10.58;$$

$$(b + h)/2t = (120 + 120)/2t = 240/24 = 10.$$

Here, h and b are the horizontal and vertical dimensions, respectively, of each angle and t is the thickness. Referring to Table 5.2 (sheet 3) of Eurocode 3, Part 1-1, for class 3 classification,

$$h/t \leq 15\varepsilon \quad \text{and} \quad (b + h)/2t \leq 11.5\varepsilon.$$

In our case, the section satisfies both conditions. So, OK.

We now consider the buckling resistance to compression. Referring to Clause 6.3.1, a compression member should be verified against buckling by the following equation:

$$N_{\mathrm{Ed}}/N_{\mathrm{b,Rd}} \leq 1.0 \tag{6.46}$$

where $N_{\mathrm{b,Rd}}$ is the design buckling resistance of the compression member and N_{Ed} is the ultimate design compressive force. The buckling resistance of a compression member is given by the following equation:

$$N_{\mathrm{b,Rd}} = \chi A f_y/\gamma_{\mathrm{M1}} \tag{6.47}$$

where χ is the reduction factor for the relevant buckling mode, A is the gross cross-sectional area = 55 cm^2, f_y = 275 N/mm^2 and γ_{M1} = partial factor = 1.0 (see Clause 6.1, note 2B). Referring to the equation

$$\bar{\lambda} = L_{\mathrm{cr}}/(i_y \lambda_1) \tag{6.50}$$

(see Eurocode 3, Part 1-1), where $\lambda_1 = 93.3\varepsilon = 86.4$, L_{cr} = effective length = 305 cm and $i_y = 3.65$,

$$\bar{\lambda} = 305/(3.65 \times 86.4) = 0.97.$$

Referring to Table 6.2 ("Selection of buckling curve for a section") of Eurocode 3, Part 1-1, in our case, we have an angle section, so we follow the buckling curve "b" in Fig. 6.4 of that Eurocode. With $\lambda^{-} = 0.97$, $\chi = 0.6$. Therefore

$$N_{b,Rd} = 0.6 \times 55 \times 100 \times 275/10^3 = 908 \text{ kN} > N_{Ed} \text{ (771 kN)};$$

$$N_{Ed}/N_{b,Rd} = 771/908 = 0.85 < 1.0. \qquad \underline{\text{OK}}$$

Therefore we adopt two angles 120 × 120 × 12 back to back with a 12 mm gap between the vertical faces for the diagonal member (5-12).

In fact, for all diagonals, we use two angles 120 × 120 × 12 back to back. For all horizontals, we use two angles 100 × 100 × 10 back to back.

Next, we consider the compression chord member (4-5).

Force N_{Ed} = 1441 kN (compression).

Try two angles 150 × 150 × 18: $A = 102 \text{ cm}^2$, $r = 4.55$ cm, $L_{cr} = 450$ cm,

$$\bar{\lambda} = L_{cr}/(i_y\lambda_1) = 450/(4.55 \times 86.4) = 1.14.$$

Referring to curve "b" in Fig. 6.4 of Eurocode 3, Part 1-1, $\chi = 0.53$. Therefore

$$N_{b,Rd} = 0.53 \times 102 \times 100 \times 275/103 = 1487 \text{ kN} > N_{Ed} \text{ (1441 kN)};$$

$$N_{Ed}/N_{b,Rd} = 1441/1487 = 0.97 < 1.0. \qquad \underline{\text{OK}}$$

Therefore we adopt two angles 150 × 150 × 18 back to back with a 12 mm gap between the vertical faces.

Therefore, for both tension and compression chord members, we adopt two angles 150 × 150 × 18 back to back (see Fig. 7.1).

7.5 Design of horizontal wind girder at 33.0 m level

7.5.1 Design considerations

The horizontal wind girder at 33.0 m level is formed by latticing the bottom chords of the roof trusses between rows 9 and 10. The wind girder takes the reactions due to wind loads from the gable end columns, which are supported at their top ends on the wind girder. Similarly, the same type of wind girder is formed between stanchion rows 1 and 2 to take the wind reactions on the gable end columns located along row 1. So, the span of the lattice girder is 27.0 m and its depth is 6.0 m (see Fig. 7.1).

7.5.2 Loadings

The reaction at the node point from the gable end columns at 33.0 m level is 5.4 kN (unfactored). Assuming a safety factor $\gamma_{wk} = 1.5$,

ultimate load at node point = 1.5 × 5.4 = 8.1, say 8 kN.

7.5.3 Calculation of forces in members of lattice girder

Consider the lattice girder (1-7-8-14) subjected to wind load reactions at node points 1, 2, 3, 4, 5, 6 and 7. From previous calculations,

wind point loads at nodes 1 and 7 = 4 kN;

wind point loads at nodes 2, 3, 4, 5 and 6 = 8 kN;

reaction at left support = $8 \times 3 = 24$ kN;

reaction at right support = 24 kN

(see Fig. 7.1).

Consider node 8.

Reactions at supports 8 and 14 = 24 kN.

Force in member (1-8) = 24 kN (compression).

Force in member (8-9) = 0,

as there is no external horizontal force.

Next, consider node 1: diagonal length $= (6^2 + 4.5^2)^{0.5} = 7.5$ m.

$\sum V = 0$: $(1\text{-}9) \times 6/7.5 + 4 - (1\text{-}8) = 0$.

Therefore

force in member (1-9) = 25 kN (tension).

$\sum H = 0$: $(1\text{-}2) = (1\text{-}9) \times 4.5/7.5 = 15$.

Therefore

force in members (1-2) and (2-3) = 15 kN (compression).

Next, consider node 9.

$\sum V = 0$: $-(1\text{-}9) \times 6/7.5 + (2\text{-}9) + (3\text{-}9) \times 6/7.5 = 0$.

Therefore

force in member (3-9) = 15 kN (compression);

force in member (2-9) = 8 kN (compression).

$\sum H = 0$: $(9\text{-}10) = (1\text{-}9) \times 4.5/7.5 + (3\text{-}9) \times 4.5/7.5$.

Therefore

force in member (9-10) = 48 kN (tension).

Next, consider node 3.

$\sum V = 0$: $-(3\text{-}9) \times 6/7.5 + (3\text{-}11) \times 6/7.5 = 0$.

Therefore

force in member (3-11) = 15 kN (tension).

$\sum H = 0$: (2-3) + (3-9) × 4.5/7.5 + (3-11) × 4.5/7.5 − (3-4) = 0.

Therefore

force in member (3-4) = 33 kN (compression).

7.5.4 Design of section of members

Consider the diagonal member (3-9). The force in this member is 15 kN (compression). Try two angles 125 × 75 × 10 (125 mm legs back to back): $A = 38.2\ \text{cm}^2$, $i_z = 3.14$ cm, $L_{cr} = 750$ cm, $l_z = 750$ cm,

$$\bar{\lambda} = 750/(i_z \lambda_1) = 750/(3.14 \times 86.4) = 2.76.$$

Referring to Table 6.2 of Eurocode 3, Part 1-1, for our L-section, we follow the buckling curve "b" in Fig. 6.4 of that Eurocode. With $\bar{\lambda} = 2.76$, $\chi = 0.1$, and

$$N_{b,Rd} = \chi A f_y / M_1 = 0.1 \times 38.2 \times 100 \times 275/10^3 = 105\ \text{kN} >> N_{Ed}\ (15\ \text{kN}).$$

Therefore we adopt two angles 125 × 75 × 10 (125 mm legs back to back).

We adopt the same section for all members. See Table 7.3 for the forces in the members and the sizes of those members.

Table 7.3. Forces in and sizes of members

Member	Location	Force (kN), ultimate	Member size, 2 angles $h \times b \times t$
1-2, 6-7	Outer chords	15 (C)	100 × 100 × 10
2-3, 5-6	Outer chords	15 (C)	100 × 100 × 10
3-4, 4-5	Outer chords	33 (C)	100 × 100 × 10
8-9, 13-14	Inner chords	0	100 × 100 × 10
9-10, 12-13	Inner chords	48 (T)	100 × 100 × 10
10-11, 11-12	Inner chords	48 (T)	100 × 100 × 10
1-9, 7-13	Diagonals	25 (T)	125 × 75 × 10
3-9, 5-13	Diagonals	15 (C)	125 × 75 × 10
3-11, 5-11	Diagonals	15 (T)	125 × 75 × 10
1-8, 7-14	Horizontals	24 (C)	125 × 75 × 10
2-9, 6-13	Horizontals	8 (C)	125 × 75 × 10
3-10, 5-12	Horizontals	0	125 × 75 × 10
4-11	Horizontal	8 (C)	125 × 75 × 10

References

Butterworth, 1949. *Structural Analysis by Moment Distribution*, Longmans, Green & Co., New York.

Eurocode, 2005. BS EN 1993-1-1: 2005, Eurocode 3. Design of steel structures. General rules and rules for buildings.

CHAPTER 8

Case Study III: Design of Vertical Bracing Systems for Wind Forces and Crane Tractive Forces Along Stanchion Lines A and B, Based on Eurocode 3

8.1 Vertical bracing systems along stanchion line A

8.1.1 Design considerations (see Fig. 1.5)

Two vertical lines of bracing systems are provided along stanchion line A, one along the crane columns and the other along the roof columns. The arrangement of the system is of an inverted V type, as shown in Fig. 1.5. This type of bracing system has been adopted to provide maximum clearance for the movement of equipment within the building. The bracing system along the crane columns ends just below crane girder level, and the other bracing system, along the roof columns, extends up to roof truss level. Both systems start from ground floor level. The bracing systems are formed between stanchion rows 9 and 10 and between rows 1 and 2 at the extreme ends of the building, and also at about the centre of the length of the building between rows 5 and 6 (see Fig. 1.5 again).

8.1.2 Functions

The main function of the vertical bracing system along the crane column line is to resist the longitudinal horizontal tractive force generated during the longitudinal movements of the cranes during operation. This force is transferred through the crane girders to the column caps. We assume that the bracing system at each end of the building takes the full tractive force. In addition, the bracing system resists the horizontal reaction from the wind girder located at 22.5 m level due to wind on the gable end.

The main function of the vertical bracing system along roof column line of stanchions is to resist the wind loads on the gable end. The location and type should be the same as in the crane column line.

8.2 Design of bracing system between crane column rows 9 and 10 along stanchion line A to resist the longitudinal tractive force due to crane loads and wind loads from the gable end (see Fig. 8.1)

We consider the bracing frame system 5-6-9-10 between crane column rows 9 and 10 (see Fig. 8.1(b)).

8.2.1 Loadings

8.2.1.1 Longitudinal tractive force

Longitudinal tractive force = 208 kN (unfactored)

(see gantry girder calculations). We assume a partial safety factor $\gamma_{crk} = 1.5$ (see BS EN 1990: 2002 (Eurocode, 2002)).

8.2.1.2 Reaction from intermediate wind girder at 22.36 m level due to wind on gable end

Horizontal reaction force due to wind = 379.7 kN

(unfactored, previously calculated in Chapter 7). In addition, there is a uniform wind load on the crane column from the gable end, equal to $p_e = 1.4$ kN/m^2. The width of the gable end sheeting wall supported by the crane column is $b = 2.25$ m (see Fig. 7.1). Therefore

wind force/m height of column $w_k = p_e b = 1.4 \times 2.25 = 3.15$ kN/m height.

So, wind force at node 5 = $3.15 \times 10.64/2 = 16.8$ kN

and wind force at node 3 = $3.15 \times 22.36/2 = 35.2$ kN

(see frame 5-6-9-10 in Fig. 8.1(b)).

8.2.1.3 Ultimate design forces

Case 1: when the crane tractive force is operating but no wind is acting. With a partial safety factor $\gamma_{crk} = 1.5$,

ultimate design tractive force = F_{cr} at node 5 (22.36 m level) = $1.5 \times 208 = 312$ kN.

Case 2: when the tractive force and wind force are acting simultaneously. Referring to BS EN 1990: 2002 (Eurocode, 2002), we have the following two subcases.

Case 2a: when the crane tractive force is considered as the leading variable and the wind force as an accompanying variable. With a partial safety factor for the crane load $\gamma_{crk} = 1.5$ and a partial safety factor for the wind load $\gamma_{wk} = 0.6 \times 1.5 = 0.9$,

ultimate design force due to (crane tractive force + wind) at node 5

$= 1.5 \times 208 + 0.9 \times (379.7 + 16.8) = 668.9$ kN.

Case 2b: when the wind force is considered as the leading variable and the crane tractive force as an accompanying variable. With a partial safety factor for the wind load $\gamma_{wk} = 1.5$ and a partial safety factor for the crane tractive force $\gamma_{crk} = 0.9$,

ultimate design force = $1.5 \times (379.7 + 16.8) + 0.9 \times 208 = 782$ kN.

So, case 2b will be considered in the analysis. Therefore, at node 5,

ultimate design force = $F_{u5} = 782$ kN,

and at node 3,

ultimate design force = $F_{u3} = 1.5 \times 35.2 = 53$ kN.

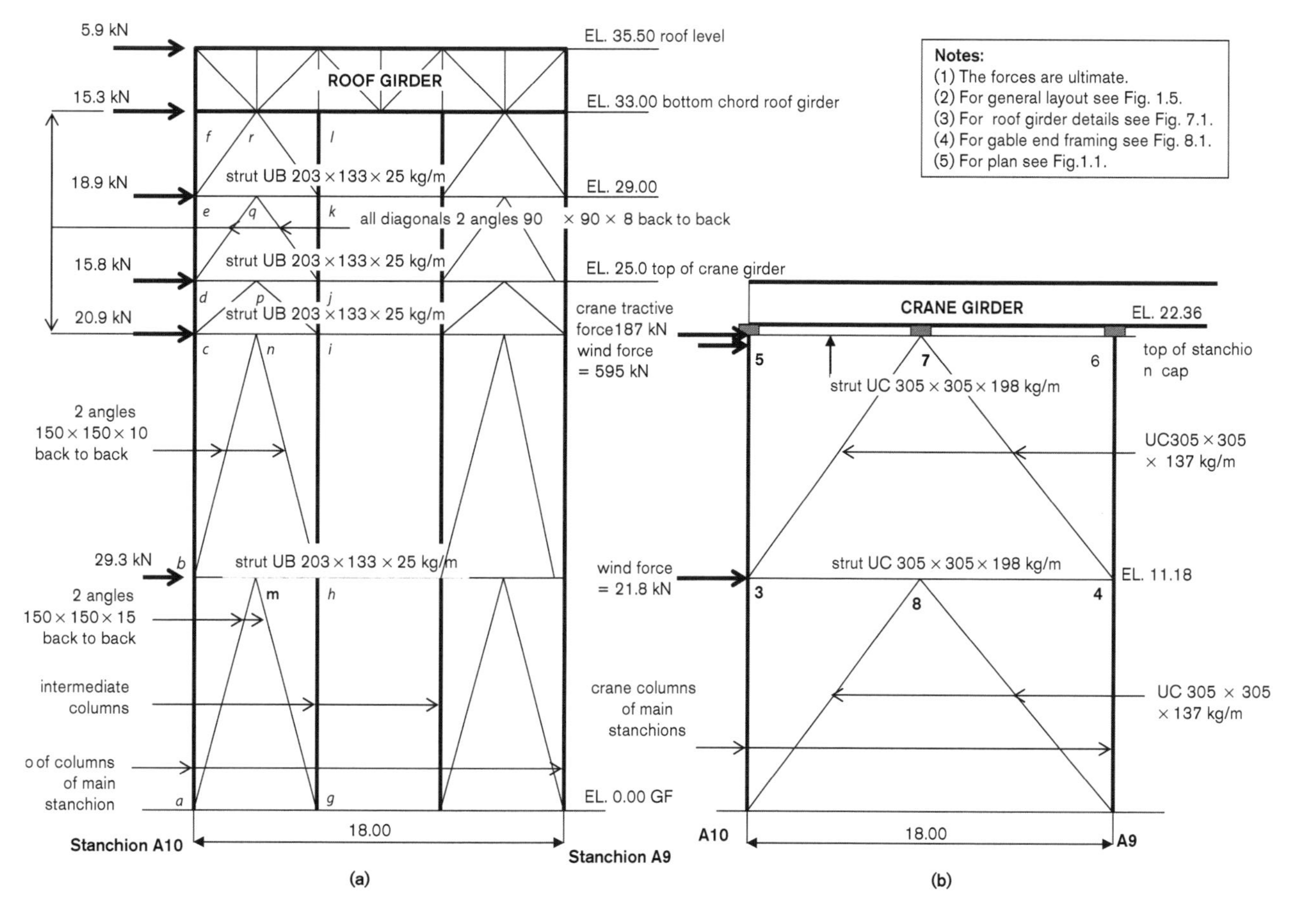

Fig. 8.1. Bracing systems for wind forces and crane tractive forces between stanchions A9 and A10. (a) Wind bracings along roof column; (b) bracings for longitudinal tractive force of crane

8.2.2 Forces in the members of the bracing system along crane column

Consider the bracing system between crane column rows 9 and 10 (see Fig. 8.1(b)).

First, we apply the ultimate design force 782 kN at node point 5 and analyse the frame. The slope of the diagonals with respect to the horizontal is θ, where

$\tan \theta = 11.18/9 = 1.24$; $\theta = 51.2°$;

$\sin 51.2° = 0.78$; $\cos 51.2° = 0.63$.

Consider node 7.

$\sum V = 0$:

$+ (4\text{-}7) \sin \theta - (3\text{-}7) \sin \theta = 0$;

$$+ (4\text{-}7) \times 0.78 - (3\text{-}7) \times 0.78 = 0. \quad (1)$$

$\sum H = 0$:

$$-(4\text{-}7) \cos \theta - (3\text{-}7) \cos \theta + 782 = 0; \quad (2)$$

$$-(4\text{-}7) \times 0.63 - (3\text{-}7) \times 0.63 + 782 = 0. \quad (2)$$

Multiply both sides by a factor 0.78/0.63 = 1.25:

$$-(4\text{-}7) \times 0.78 - (3\text{-}7) \times 0.78 + 977.5 = 0. \quad (3)$$

Add (1) and (2):

$1.56 \times (3\text{-}7) = 977.5$.

Therefore

force in diagonal member (3-7) = 977.5/1.56 = 627 kN (tension).

Insert this value in equation (1):

$(4\text{-}7) \times 0.78 - 0.78 \times 627 = 0$.

Therefore

force in diagonal member (4-7) = 0.78 × 627/0.78 = 627 kN (compression);

force in horizontal member (5-7) = 782 kN (compression).

Next, we apply the ultimate force (782 + 53) = 835 at node point node 3.

Force in diagonal member (8-9) = 835/2 × 14.35/9 = 661 kN (compression);

force in diagonal member (8-10) = 661 (tension);

force in horizontal member (3-9) = 835 kN (compression).

8.2.3 Design of section of members, based on Eurocode 3, Part 1-1 (Eurocode, 2002)

8.2.3.1 Diagonal member (8-9)

Maximum ultimate design force in diagonal member (8-9) = 661 kN (compression).

Try UC305 × 305 × 137 kg/m: $A = 174$ cm^2, $iz = 7.83$ cm, $L_{crz} = (11.18^2 + 9.0^2)^{0.5} = 14.35$ m.

$\lambda_1 = 86.4$ (previously calculated).

$\bar{\lambda} = L_{crz}/(i_z\lambda_1) = 14.35 \times 100/(7.83 \times 86.4) = 2.12.$

Referring to Table 6.2 of Eurocode 3, Part 1-1, for a rolled section with $h/b = 320.5/309.2 = 1.04 \leq 1.2$ and $t_f \leq 100$, we follow curve "c" for the axis z–z and find that $\chi = 0182$. Therefore

$N_{b,Rd} = \chi A f_y/\gamma_{M1} = 0182 \times 174 \times 275/10 = 861$ kN >> N_{Ed} (661 kN);

$N_{Ed}/N_{b,Rd} = 661/861 = 0.77 < 1.$ OK

Therefore we adopt UC 305 × 305 × 137 kg/m.

8.2.3.2 Horizontal member (3-8)

Maximum ultimate design force in horizontal member (3-8) = 835 kN (compression).

Try UC305 × 305 × 198 kg/m: $A = 252$ cm^2, $i_z = 8.04$ cm, $L_{cr} = 1800$ cm.

$\lambda_1 = 86.4$ (previously calculated).

$\bar{\lambda} = L_{crz}/(i_z\lambda_1) = 1800/(8.04 \times 86.4) = 2.59.$

Referring to Table 6.2 of Eurocode 3, Part 1-1, for a rolled section with $h/b = 339.9/314.5 = 1.08 \leq 1.2$ and $t_f \leq 100$ mm, we follow curve "c" for the axis z–z in Fig. 6.4 of that Eurocode and find that $\chi = 0.12$.
Therefore

$N_{b,Rd} = \chi A f_y/\gamma_{M1} = 0.12 \times 252 \times 275/10 = 832$ kN < 835 kN,

but we accept this. Therefore, we adopt UC305 × 305 × 198 kg/m.

For the horizontal member (5-6), we adopt UC305 × 305 × 198 kg/m.

For all diagonals, we adopt UC305 × 305 × 137 kg/m.

8.3 Design of vertical bracing system between roof column rows 9 and 10 along stanchion line A to resist wind loads from gable end (see Fig. 8.1(a))

8.3.1 Design considerations

The bracing system should ideally be arranged between roof column rows 9 and 10, spaced 18 m apart. But intermediate columns are located between the stanchions at a spacing of

6.0 m. So, the bracing system is arranged between the stanchions in roof column row 10 and the intermediate columns at a distance of 6.0 m, as shown in Fig. 8.1(a). The wind loads from the gable end are applied at the node points.

8.3.2 Loadings

We refer to Fig. 8.1(a) and also the gable end shown in Fig. 7.1. The wind load/m height w_{k1} from ground level to 22.36 m level is given by $p_e b_1$ (where b_1 is half the spacing between the crane column and the roof column) = 1.4 × 2.5/2 = 1.75 kN/m height. The wind load/m height w_{k2} from 22.36 m level to 35.5 m level is again given by $p_e b_1$ (where b_1 is now half the spacing between the roof column and the gable column) = 1.4 × 4.5/2 = 3.15 kN/m. With a partial safety factor for wind loading $\gamma_{wk} = 1.5$, the ultimate wind load at the node point at 11.25 m level is

$P_1 = w_{k1} \times$ (half the height from ground level to stanchion cap) × 1.5
= 1.75 × 11.18 × 1.5 = 29.3 kN.

Ultimate wind load at node point at 22.36 m level = P_2= [1.75 × 11.18/2 + 3.15 × 2.64/2] × 1.5 = 20.9 kN.

Ultimate wind load at node point at 25.0 m level = P_3 = [3.15 × 3.32] × 1.5 = 15.8 kN.

Ultimate wind load at node point at 29.0 m level = P_4 = 3.15 × 4 × 1.5 = 18.9 kN.

Ultimate wind load at node point at 33.0 m level = P_5 = 3.15 × 3.25 × 1.5 = 15.3 kN.

Ultimate wind load at node point at 35.5 m level = P_6 = 3.15 × 1.25 × 1.5 = 5.9 kN.

8.3.3 Forces in members

Ultimate total shear at ground floor level

$= V_{u1} = P_1 + P_2 + P_3 + P_4 + P_5 + P_6 = (29.3 + 20.9 + 15.8 + 18.9 + 15.3 + 5.9)$ kN = 106.1 kN.

Ultimate design shear V_{u2} at 11.18 m level = (106.1 – 29.3) kN = 76.8 kN.

Ultimate design force V_{u3} at 22.36 m level = (76.8 – 20.9) = 55.9 kN.

Ultimate design force V_{u4} at 25.0 m level = (55.9 – 15.8) = 40.1 kN.

Ultimate design force F_{u5} at 29.0 m level = (40.1 – 18.9) = 21.2 kN.

Ultimate design force F_{u6} at 33.0 m level = (21.2 – 15.3) = 5.9 kN.

Consider the diagonal member (g-m) at ground floor level.
Length of diagonal = $(11.25^2 + 3^2)^{0.5}$ = 11.72 m.

Force in diagonal (g-m) = 106.1/2 × 11.72/3 = 207 kN (compression).

Force in diagonal (a-m) = 207 kN (tension).

Next, consider the diagonal member (h-n) at 11.18 m level.

Force in diagonal (h-n) = 76.8/2 × 11.72/3 = 143 kN (compression).

Force in diagonal (b-n) = 143 kN (tension).

Next, consider the diagonal member (i-p) at 22.36 m level.

Length of diagonal = $(3^3 + 2.64^2)^{0.5} = 4$ m.

Force in diagonal (i-p) = $55.9/2 \times 4/3 = 37$ kN (compression).

Force in diagonal (c-p) = 37 kN (tension).

Consider the diagonal member (j-q) at 25 m level:

Length of diagonal = $(4^2 + 3^2) = 5$ m.

Force in diagonal (j-q) = $40.1/2 \times 5/3 = 33.0$ kN (compression).

Force in diagonal (d-q) = 33 kN (tension).

Consider the diagonal member (k-r) at 29 m level.

Force in diagonal (k-r) = $21.2/2 \times 5/3 = 18$ kN (compression).

Force in diagonal (e-r) = 18 kN (tension).

8.3.4 Design of section of members

Consider a diagonal member at ground floor level.

Force in the member = 207 kN (compression).

Try two angles 150 × 150 × 15: $A = 86$ cm^2; $r_y = 4.57$ cm, $L_{cry} = 1172$ cm,

$$\lambda_1 = 86.4,$$

$$\bar{\lambda} = L_{cr}/(\lambda_1 i_y) = 1172/(86.4 \times 4.57) = 2.97.$$

Referring to Table 6.2 of Eurocode 3, Part 1-1, for an angle cross-section, we follow the buckling curve "b" in Fig. 6.4 of that Eurocode. With $\bar{\lambda} = 2.97$, $\chi = 0.1$. Therefore

$$N_{b,Rd} = \chi A f_y / \gamma_{M1} = 0.1 \times 86 \times 275/10 = 236 \text{ kN} > N_{Ed} \text{ (207 kN)}. \qquad \underline{\text{OK}}$$

Therefore we adopt two angles 150 × 150 × 15 back to back for the diagonal members at ground floor level.

We adopt the same section for the diagonals between 11.18 m level and 22.36 m level. Next, consider a diagonal at 25.0 m level.

Force in the diagonal = 33 kN (compression).

Try two angles 90 × 90 × 8 back to back: $A = 27.8$ cm^2, $iy = 2.74$ cm, $L_{cry} = 5$ m,

$$\bar{\lambda} = L_{cr}/(\lambda_1 iy) = 500/(86.4 \times 2.74) = 2.1.$$

Referring to Table 6.2 of Eurocode 3, Part 1-1, for an angle cross-section, we follow the buckling curve "b" in Fig. 6.4 of that Eurocode. With $\bar{\lambda} = 2.1$, $\chi = 0.18$. Therefore

$$N_{b,Rd} = 0.18 \times 27.8 \times 275/10 = 137 \text{ kN} > N_{Ed} \text{ (33 kN)}. \qquad \underline{\text{OK}}$$

Therefore we adopt two angles 90 × 90 × 8 back to back from 22.36 m level to 33.0 m level. See Table 8.1 for the forces in the members and the sizes of those members.

Table 8.1. Forces in and sizes of members

Member	Location	Force (kN), ultimate[a]	Member size
(a-m), (g-m)	Diagonals between ground level and 11.18 m	Member (a-m), 207 (T) Member (g-m), 207 (C)	2 angles 150 × 150 × 15 2 angles 150 × 150 × 15
(b-n), (h-n)	Diagonals between 11.18 m and 22.36 m	Member (b-n), 143 (T) Member (h-n), 143 (C)	2 angles 150 × 150 × 15 2 angles 150 × 150 × 15
(c-p), (p-l)	Diagonals between 22.36 m and 25.0 m	Member (c-p), 37 (T) Member (l-p), 37 (C)	2 angles 90 × 90 × 8 2 angles 90 × 90 × 8
(d-q), (j-q)	Diagonals between 25.0 m and 29.0 m	Member (d-q), 33 (T) Member (j-q), 33 (C)	2 angles 90 × 90 × 8 2 angles 90 × 90 × 8
(e-r), (k-r)	Diagonals between 29.0 m and 33.0 m	Member (e-r), 18 (T) Member (k-r), 18 (C)	2 angles 90 × 90 × 8 2 angles 90 × 90 × 8
(b-m), (c-n), (d-p), (e-q)	Horizontals	Member (b-m), 80.4 (C) Member (c-n), 51.1 (C)	UB203 × 133 × 25 kg/m UB203 × 133 × 25 kg/m

[a] Compression forces are denoted by (C) and tension forces by (T).

8.4 Design of vertical bracing system for wind forces and crane tractive forces in column along stanchion line B

8.4.1 Design considerations

The wind bracing system is subjected to reactions from wind girders and direct uniform loads due to wind on the gable end (see Fig. 7.1). To allow the free movement of equipment and vehicles, no bracings could be provided up to the stanchion cap level (22.5 m) (see Fig. 8.2). So, a beam was placed at 22.50 m level, with moment connections at the ends to the columns to resist the moments due to wind forces.

8.4.2 Wind loadings (see Fig. 7.1)

8.4.2.1 Characteristic horizontal forces at node points due to wind and tractive force

Force at 43.5 m level due to uniform wind pressure on column

$= P_1 = w_k a_1 = 1.4 \times 7.25 \times 3 = 30.5$ kN,

where w_k is the wind pressure on the gable wall = 1.4 kN/m^2 (already calculated) and a_1 is the area of gable wall supported by the column.

Reaction from wind girder at 43.5 m level $= P_2 = w_k a_2 = 1.4 \times 7.25/2 \times 6 = 30.5$ kN.

Force at 33.0 m level due to uniform wind pressure on column

$= P_3 = w_k a_3 = 1.4 \times (10.57 \times 3 + 7.82 \times 2.25) = 69$ kN.

Reaction from wind girder at 33.0 m level $= P_4 = w_k a_4 = 1.4 \times 10.57 \times 6/2 = 44.4$ kN.

Reaction from wind girder at 33.0 m level for a span of 27 m $= P_5 = 3 \times 5.4 = 16$ kN.

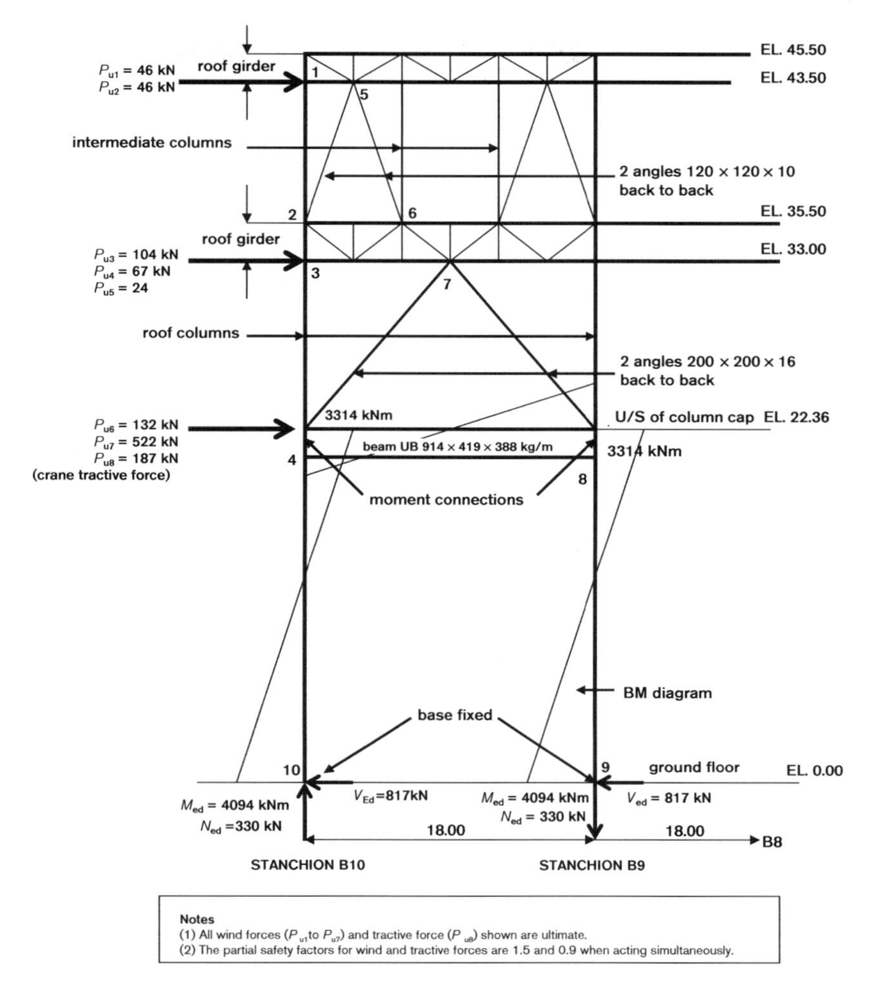

Fig. 8.2. Vertical bracing system for wind and tractive forces along stanchion line B

Force at 22.36 m level due to uniform wind pressure on column

$= P_6 = w_k a_5 = 1.4 \times (10.64/2 \times 5.5 + 22.36/2 \times 3) = 87.9$ kN.

Reaction from wind girder at 22.36 m level $= P_7 = 522/1.5 = 348$ kN.

Crane longitudinal tractive force at 22.36 m level $= P_8 = 208$ kN.

All of the above forces are unfactored.

8.4.2.2 Ultimate design horizontal forces at node points

With a partial safety factor $\gamma_{wk} = 1.5$ for wind loads (as previously explained), the ultimate design forces are

$P_{u1} = 1.5 \times 30.5 = 46$ kN, $P_{u2} = 1.5 \times 30.5 = 46$ kN,

$P_{u3} = 1.5 \times 69 = 104$ kN, $P_{u4} = 15 \times 44.4 = 67$ kN,

$P_{u5} = 1.5 \times 16 = 24$ kN, $P_{u6} = 15 \times 87.9 = 132$ kN,

$P_{u7} = 1.5 \times 348 = 522$ kN, $P_{u8} = 0.6 \times 1.5 \times 208 = 187$ kN (see below).

The wind forces are assumed to be the main variable forces, with a partial safety factor $\gamma_{wk} = 1.5$, and the crane tractive force is considered as an accompanying variable force, with $\gamma_{crk} = 0.9 \times 1.5 = 0.9$ (see Fig. 8.2).

8.4.3 Analysis of frame

Ultimate total wind shear at ground level V_{u1}

$= P_{u1} + P_{u2} + P_{u3} + P_{u4} + P_{u5} + P_{u6} + P_{u7}$
$= 46 + 46 + 104 + 67 + 24 + 132 + 522 = 941$ kN.

Ultimate total wind shear at 22.36 m level V_{u2}

$= P_{u1} + P_{u2} + P_{u3} + P_{u4} + P_{u5} = 46 + 46 + 104 + 67 + 24 = 287$ kN.

Ultimate total shear at 33.0 m level $V_{u3} = P_{u1} + P_{u2} = 46 + 46 = 92$ kN.

We apply the above ultimate design wind loads at node points and analyse the frame.

Ultimate shear at ground level due to tractive force $= V_{u4} = 0.9 \times 208 = 187$ kN

(acting simultaneously with wind).

We apply the above ultimate wind shear forces at node points at 22.36 m level and above, and analyse the bracing system without the crane tractive force.

Length of diagonal (7-8) $= (10.64^2 + 9^2)^{0.5} = 13.94$ m.

Consider the total ultimate shear V_{u2} at level 22.36 m = 287 kN.

Force in diagonal (7-8) $= 287/2 \times 13.94/9 = 222$ kN (compression).

Force in diagonal (4-7) = 222 kN (tension).

Force in horizontal (3-7) = 287 kN (compression).

Next, consider the total shear V_{u3} at level 33.0 m = 92 kN. We assume that the first two diagonal bracings take the whole shear.

Length of diagonal $= (8^2 + 3^2)^{0.5} = 8.54$ m.

Force in diagonal (5-6) $= 92/2 \times 8.54/3 = 131$ kN (compression).

Force in diagonal (2-5) = 131 kN (tension).

Force in horizontal (1-5) = 92 kN (compression).

Now, we consider (4-8-9-10) as a portal frame with moment connections at joints 4 and 8 and fixed at the base at joints 9 and 10. The frame is subjected to an ultimate horizontal design load due to wind at node 4,

$$V_{u1} = P_{u1} + P_{u1} + P_{u2} + P_{u3} + P_{u4} + P_{u5} + P_{u6} + P_{u7} = 941 \text{ kN},$$

and an ultimate design force due to the tractive force $P_{u8} = 187$ kN.

We analyse the frame (4-8-9-10) as follows. The stanchions along line B are each made up of two columns, one carrying the gantry girder reaction load and half of the upper-roof-column load at level 22.36 m. The other, lower, roof column carries half of the upper-roof-column load at level 22.36 m. The crane tractive force is carried by only the crane columns. This should be clear from the details of the stanchions in line B shown in Fig. 6.6. From the arrangement shown in Fig. 6.6, we see that the frame (4-8-9-10) is composed of two framing systems, one along the crane column and the other along the lower roof column. Thus, we can assume that the total ultimate horizontal wind load (shear) of 941 kN at node 4 is shared equally by these two framing systems. Therefore, the crane column at 22.36 m level will carry a shear $V_u = V_{u1}/2 + V_{u4} = 941/2 + 187 = 658$ kN and the lower roof column at 22.36 m level will carry $V_{u1}/2 = 471$ kN.

To analyse the frame, we consider the frame (4-8-9-10) along the crane column line, subjected to a horizontal shear of 658 kN at node 4, at 22.36 m level. We assume a section of column UB914 × 419 × 388 kg/m (previously adopted for the crane columns in stanchion line B) and a section of beam UB914 × 419 × 388 kg/m. We assume the bases are fixed.

$I_y = 720\,000$ cm^4, $i_y = 38.2$ cm, $i_z = 8.59$ cm.

Height of column $= h = 22.36$ m; length of beam $= l = 18$ m.

Stiffness of column $= K(9\text{-}8) = K(4\text{-}10) = I_y/h = 720\,000/(22.5 \times 100) = 320$.

Stiffness of beam $= K(4\text{-}8) = I_y/l = 720\,000/(18 \times 100) = 400$.

The distribution factors for joint 4 are as follows.

Distribution factor $DF(4\text{-}10)$ for column $= K(4\text{-}10)/[K(4\text{-}10) + K(4\text{-}8)]$
$= 320/(320 + 400) = 0.44$.

Distribution factor $DF(4\text{-}8) = 0.56$.

We assume arbitrary fixed end moments in columns (4-9) and (8-10) equal to $M_f(4\text{-}9) = M_f(8\text{-}10) = 100$ kN m (since $-M_f(4\text{-}9) : -M_f(8\text{-}10) = K(4\text{-}9) : K(8\text{-}10)$ for columns of equal length). With these arbitrary fixed end moments of 100 kN m, we carry out a moment distribution analysis of the frame, as shown in Table 8.2. The portal frame (10-4-8-9) is "flattened out" so that it becomes a continuous beam.

From the final moments,

horizontal reaction at base $H_9 = H_{10} = -(84 + 68)/22.5 = -6.75$ kN.

To induce equilibrium,

$X = -(H_9 + H_{10}) = +2 \times 6.75 = +13.5$ kN.

The positive sign indicates that X acts from left to right. Because the actual sway force is 658 kN (previously calculated (wind + crane tractive) shear force at node 4), the moments resulting from the moment distribution analysis should be multiplied by a factor 658/13.5 = 48.74. Therefore the final design moments are as follows:

- at node 9, –84 × 48.74 = 4094 kN m;
- at node 4, –68 × 48.74 = 3314 kN m;
- at node 8, –3314 kN m;

Table 8.2. Moment distribution analysis by Hardy Cross method (Fisher Cassie, 1951)

	Fixed base					Fixed base
	10	Column 4	Beam		8 Column	9
Distribution factor		0.44	0.56	0.56	0.44	
	−100	−100			−100	
		+44	+56	+56	+44	
	+22		+28			
		−12	−16			
	−6					
	−84	−68	+68	+68	−68	−84

- at node 10, −4094 kN m.

Shear at node 9 = −(4094 + 3314)/22.36 = −330 kN and at 10 = −330 kN.

Therefore, for the beam (4-8),

ultimate design moment M_{Ed} = 3314 kN m;

ultimate design shear $V_{Ed} = 658 \times 22.26/18 = 817$ kN;

ultimate design thrust N_{Ed} = 330 kN.

The BM diagram, shear and thrust are shown in Fig. 8.2.

8.4.4 Design of section of members

8.4.4.1 Diagonal member (5-6)

Force in member = 131 kN (compression).

Try two angles 120 × 120 × 10: i_y = 3.67 cm, A = 46.4 cm^2, L_{cr} = 854 cm.

$\lambda_1 = 86.4$,

$\bar{\lambda} = L_{cr}/(i_y\lambda_1) = 854/(3.67 \times 86.4) = 2.69$.

Referring to Fig. 6.4 of Eurocode 3, Part 1-1, $\chi = 0.1$. Therefore

$N_{b,Rd} = 0.1 \times 46.4 \times 100 \times 275/10^3 = 128 \text{ kN} > N_{Ed}$ (131 kN).

Therefore we adopt two angles 120 × 120 × 10 back to back.

8.4.4.2 Diagonal member (7-8)

Force in member = 222 kN.

Try two angles 200 × 200 × 16: $i_y = 6.16$ cm, $A = 124$ cm^2, $L_{cr} = 1380$ cm.

$\lambda_1 = 86.4$,

$\bar{\lambda} = L_{cr}/(i_y\lambda_1) = 1380/(6.16 \times 86.4) = 2.59$.

Referring to Fig. 6.4 of Eurocode 3, Part 1-1, $\chi = 0.13$. Therefore

$N_{b,Rd} = 0.13 \times 124 \times 275/103 = 443$ kN > N_{Ed} (222 kN). OK

Therefore we adopt two angles 200 × 200 × 16 back to back.

8.4.4.3 Horizontal member (4-8) of portal frame (4-8-9-10) at 22.36 m level

This member is subjected to moment, thrust and shear. So, we have to design the member taking into account all of the values given below:

- support moment $M_u = M_{Ed} = 3314$ kN m;
- thrust $N_{Ed} = 330$ kN;
- shear $V_{Ed} = 817$ kN.

The member is subjected to moment and thrust simultaneously. The section design is carried out with the above values in the following sequence.

Step 1: Selection of section, and properties of the section

Try a section UB914 × 419 × 388 kg/m: $A = 494$ cm^2, $r_y = 38.2$ cm, $r_z = 9.59$ cm, $W_y =$ 17 700 cm^3. We assume steel grade S 275, design strength $f_y = 275$ N/mm^2.

Step 2: Classification of cross-section

Before we design the section, we have to classify the section into one of the following classes:

- *Class 1, plastic:* cross-section with plastic hinge rotation capacity.
- *Class 2, compact:* cross-section with plastic moment capacity.
- *Class 3, semi-compact:* cross-section in which the stress in the extreme compression fibre reaches the design strength, but a plastic moment capacity cannot be developed.
- *Class 4, slender:* cross-section in which we have to have a special allowance owing to local buckling effects.

The thickness of the flange t_f is 36.6 mm < 40 mm. So, $f_y = 275$ N/mm^2. Therefore $\varepsilon = (235/f_y)^{0.5} = (235/275)^{0.5} = 0.92$. For a class 1 compact section, referring to Table 5.2 of Eurocode 3, Part 1-1, the limiting value of c/t_f for the flange must be less than or equal to 9ε, and the limiting value of d/t_w for a web subject to bending and compression must be less than or equal to $396\varepsilon/(13\alpha - 1)$, assuming $\alpha > 0.5$.

In our case, $9\varepsilon = 9 \times 0.92 = 8.28$ for the flange, and

$c/t_f = (b - t_w - 2r_1)/2/23.9 = (304.1 - 15.9 - 2 \times 19.1)/2/23.9 = 5.2 < 8.28$.

This is satisfactory. For the web,

$396 \times 0.92/(13 \times 0.5 - 1) = 66.24$

and

$d/t_w = 799.6/21.4 = 37.4 < 66.24$ <u>Satisfactory</u>

Thus, the selected section UB914 × 419 × 388 kg/m satisfies the criteria for a class 1 compact section.

Step 3: To check shear capacity of section

Ultimate shear force $V_{Ed} = 817$ kN.

Shear capacity of section $= V_{pl,Rd} = A_v(f_y/\sqrt{3})/\gamma_{Mo}$.

For a rolled I section,

$$A_v = A + (t_w + 2r)t_f - 2b_{tf}$$

$$= 494 + (2.14 + 2 \times 2.41) \times 3.66 - 2 \times 42.05 \times 3.66 = 211.7 \text{ cm}^2.$$

This should not be less than $\eta h_w t_w$, where $\eta = 1.0$ is recommended. We have

$\eta h_w t_w = 84.78 \times 2.14 = 181.4.$ <u>OK</u>

Therefore

$V_{pl,Rd} = 211.7 \times 100 \times (275/\sqrt{3})/10^3 = 3361 \text{ kN} >> V_{Ed}$ (817 kN). <u>Satisfactory</u>

Step 4: To check for shear buckling of web

If the ratio h_w/t_w exceeds 72ε, then the web should be checked for shear buckling. In our case,

$$d/t_w \leq 396\varepsilon/(13\alpha - 1);\ d/t_w = 799.6/21.4 = 37.4 < 396\varepsilon/(13\alpha - 1) < 66.24.$$

Therefore the web need not be checked for shear buckling.

Step 5: To check moment capacity of section

For *low shear*, if the ultimate design shear V_{Ed} does not exceed 50% of the shear capacity $V_{pl,rd}$ of the section, the moment capacity (moment resistance) $M_{b,Rd}$ of the section is equal to $f_y W_{ply}/\gamma_{Mo}$ for class 1 and class 2 compact sections.

For *high shear*, if $V_{Ed} > 0.5V_{pl,Rd}$, the resistance moment should be reduced by a factor $(1 - \rho)$ (see equation (6.29) of Eurocode 3, Part 1-1), where

$$\rho = [2 \times (V_{Ed}/V_{pl,Rd}) - 1]^2.$$

In our case, $V_{Ed} = 817$ kN, and 50% of $V_{pl,Rd}$ is 3361 kN/2 = 1680 kN. So, $V_{Ed} <$ 50% of $V_{pl,Rd}$. So, no reduction of the resistance moment is necessary.

Therefore the moment capacity is

$M_{pl,Rd} = f_y W_{pl,y}/\gamma_{Mo} = 275 \times 17\,700/10^3 = 4868 \text{ kN m} > (M_{Ed})\ 3314$ kN m. <u>OK</u>

Step 6: To check for lateral–torsional buckling

The horizontal member (4-8), as part of a portal frame, runs between crane columns. Owing to the end moments, the top flange is under compression at the near end (node 4), and the bottom flange is under compression at the far end (node 8). At midspan, there is practically no moment due to wind forces and the crane tractive force (see Fig. 8.2). Similarly, there is another beam running between the lower roof columns to form a portal frame to resist the wind force only.

The beams in these two side-by-side portal frames, 3.0 m apart, are latticed together by lacings at 45° at the top and bottom flanges. So, we may assume that the compression flange is restrained at 3.0 m intervals. So, the effective length parallel to the minor axis is $L_{crz} = L/6 = 18/6 = 3$ m.

Referring to Clause 6.3.2.4 of Eurocode 3, Part 1-1, members with discrete lateral restraint to the compression flange are not susceptible to lateral–torsional buckling if the length L_c between restraints or the resulting slenderness $\bar{\lambda}_f$ of the equivalent compression flange satisfies the following equation:

$$\bar{\lambda}_f = k_c L_c/(i_{f,z}\lambda_1) \leq \bar{\lambda}_{c,0}(M_{c,Rd}/M_{y,Ed}) \tag{6.59}$$

where $M_{y,Ed}$ is the maximum design bending moment within the restraint spacing, equal to 3314 kN m;

$$M_{c,Rd} = W_y f_y/\gamma_{M1} = 17\,700 \times 275/10^3 = 4868 \text{ kN m;}$$

L_c is the length between restraints, equal to 300 cm; k_c is the slenderness correction factor for the moment distribution within the restraint spacing, equal to 0.91 (see Table 6.6 of Eurocode 3, Part 1-1); $i_{f,z}$ is the radius of gyration of the compression flange, composed of compression flange + 1/3 of compression part of web = 9.59 cm (web neglected);

$$\bar{\lambda}_{c,0} = \bar{\lambda}_{LT,0} + 0.1 = 0.4 + 0.1 = 0.5$$

(see Clause 6.3.2.3 of Eurocode 3, Part 1-1); and

$$\lambda_1 = 93.9\varepsilon = 86.4.$$

Therefore

$$\bar{\lambda}_f = k_c L_c/(i_{f,z}\lambda_1) = 0.91 \times 300/(9.59 \times 86.4) = 0.33,$$

$$\bar{\lambda}_{c,0}(M_{c,Rd}/M_{y,Ed}) = 0.5 \times 4868/3314 = 0.73 > 0.33.$$

Thus, no reduction of buckling resistance moment will be considered.

Now,

$$M_{Ed}/M_{b,Rd} = 3314/4868 = 0.68.$$

In addition, the member is subjected to a thrust $N_{Ed} = 330$ kN;

$$N_{b,Rd} = \chi f_y A/\gamma_{M1} = 0.94 \times 275 \times 494 \times 100/10^3 = 42\,566 \text{ kN;}$$

$$N_{Ed}/N_{b,Rd} = 330/42\,566 = 0.08.$$

Table 8.3. Forces in and sizes of members

Member	Location	Force (kN), ultimate	Size
(4–7)	Diagonal at 22.36 m level	222.0 (T)	2 angles 200 × 200 × 16
(7–8)	Diagonal at 22.36 m level	222.0 (C)	2 angles 200 × 200 × 16
(3–7)	Horizontal at 33.0 m level	287.0 (C)	2 angles 200 × 200 × 20
(2–5)	Diagonal at 35.5 m level	131.0 (T)	2 angles 120 × 120 × 10
(5–6)	Diagonal at 35.5 m level	131.0 (C)	2 angles 120 × 120 × 10
(1–5)	Horizontal at 43.5 m level	92.0 (C)	2 angles 120 × 120 × 10
(4–8)	Horizontal at 22.36 m level	330.0 (C)	UB914 × 419 × 388 kg/m
		M_{Ed} at joint = 3314 kN m	
		V_{Ed} at support = 817 kN	

Therefore

$$N_{Ed}/N_{b,Rd} + M_{Ed}/M_{b,Rd} = 0.08 + 0.68 = 0.76 < 1.0.$$

Therefore, we adopt UB914 × 419 × 388 kg/m for the beam (4-8).

See Table 8.3 for the forces in the members of the frame and the sizes of those members.

References

Eurocode, 2002. BS EN 1990: 2002(E), Basis of structural design.

Eurocode, 2005. BS EN 1993-1-1: 2005, Eurocode 3. Design of steel structures. General rules and rules for buildings.

Fisher Cassie, W., 1951. *Structural Analysis: The Solution of Statically Indeterminate Structures*, Longmans Green & Co., London.

APPENDIX A

Design of Bearings of Gantry Girder

A.1 Design considerations

The crane girder is simply supported at both ends on stanchions. One end of the support is provided by a rocker bearing (hinged), and the other end by a roller bearing. As the crane girder supports very heavy moving loads from the crane, the ends are subjected to rotational movement due to deflection of the crane girder. In order to obtain rotational movement at the ends of the girder, the bearings should have curved surfaces of either cylindrical or spherical form.

The *roller bearing* consists of a bottom plate with a curved surface connected to the cap of the stanchion, and a top plate with a plane surface connected to the underside of the girder at the end to permit longitudinal translation (movement) and uniaxial rotational movement of the end of the girder over the curved surface of the bottom plate.

In the case of a *rocker bearing*, it is the usual practice to restrain longitudinal movement at one end. So, a dowel pin was inserted into the rocker bearing here to allow only rotational movement of the end to take place and to allow no longitudinal movement.

In roller bearings, a form of guide plate is connected to both of the sliding surfaces to ensure uniform sliding over the curved surface (Lee, 1990).

A.2 Material properties

The bearings are subjected to very heavy loads from gantry girder reactions due to the movement of the heavy-duty 290 t melting shop crane. The curved surface and the sliding surface of the plane plate have to withstand enormous stress-induced deformation. The ability to withstand such deformation is dependent upon the hardness of the material. A relationship exists between hardness and the ultimate strength of a material; the hardness must be not just superficial, but must relate to deep penetration into the body of the material.

Various types of material are used to cope with heavy loadings on bearings, as described below.

A.2.1 Steel

In order to achieve a high degree of hardness to resist the load on a bearing, steel grades of higher tensile strength are used. Referring to Table 3.1 of Eurocode 3, Part 1-1, "Nominal values of yield strength f_y and ultimate tensile strength f_u for hot rolled structural steel", using the high-grade steel S 460,

design strength $f_y = 460$ N/mm^2 for $t \leq 40$; $f_y = 440$ N/mm^2 for $t \leq 80$.

A.2.2 Elastomeric bearings

These are laminated bearings comprising one or more slabs of elastomer bonded to reinforcing metal plates in sandwich form to achieve a high bearing capacity. Translational movement is accommodated by shear in the elastomer: one surface of the bearing moves relative to the other. Rotational movement is accommodated by variation in the compressive strain across the elastomer. The elastomer may be either natural or synthetic rubber.

Elastomeric bearings made from the synthetic rubber "neoprene", manufactured by E.I. du Pont de Nemours and Company, Wilmington, Delaware, USA, are widely used in various highway and other projects. Neoprene has the following advantageous properties:

- High strength and physical durability: hardness in durometer test A (ASTM D 2240) = 50 to 70 (hardness grade 50–70).
- Tensile strength (ASTM D 412) = 2500 lb/inch^2 = 17.6 N/mm^2.
- Elongation at break (ASTM D 412) = 400% to 300% (hardness 50–60).
- Tear resistance (ASTM D 624) = 225 lb/linear inch = 40 N/linear mm.
- Low-temperature stiffness: maximum Young's modulus at –40 °F = 70 N/mm^2.
- It is resistant to oils and chemicals.
- It has excellent weather-ageing characteristics.
- It does not propagate flame.
- It does not embrittle in cold weather.
- It absorbs extremes of movement.
- It has excellent shock and vibration damping.

A.2.3 Conclusion

Elastomeric bearings are generally used in bridges, where the repetition of the stresses is frequent but the intensity of the stresses is low. In bearings used with crane girders, however, vertical impacts due to heavy crane loads may cause severe fatigue stresses in the material, resulting in failure of the bearings. Considering the above points, we conclude that steel bearings of higher tensile strength are best suited to resisting very heavy impact loads from cranes.

A.3 Design of bearings

We assume steel of grade S 460 and that the thickness of the bearing plate is limited to 60 mm, so that $f_y = 440$ N/mm^2. We apply the Hertz theory. See Fig. A.1.

Case 1: for a cylindrical roller of radius R_2 on a concave surface of radius R_1, the maximum contact stress is given by

$$\sigma_u = 0.418[pE(R_1 - R_2)/(R_2R_1)]^{0.5}$$

(Grinter, 1961), where

p = load per unit length of roller,

E = modulus of elasticity = 210 000 N/mm^2,

R_1 = radius of concave surface,

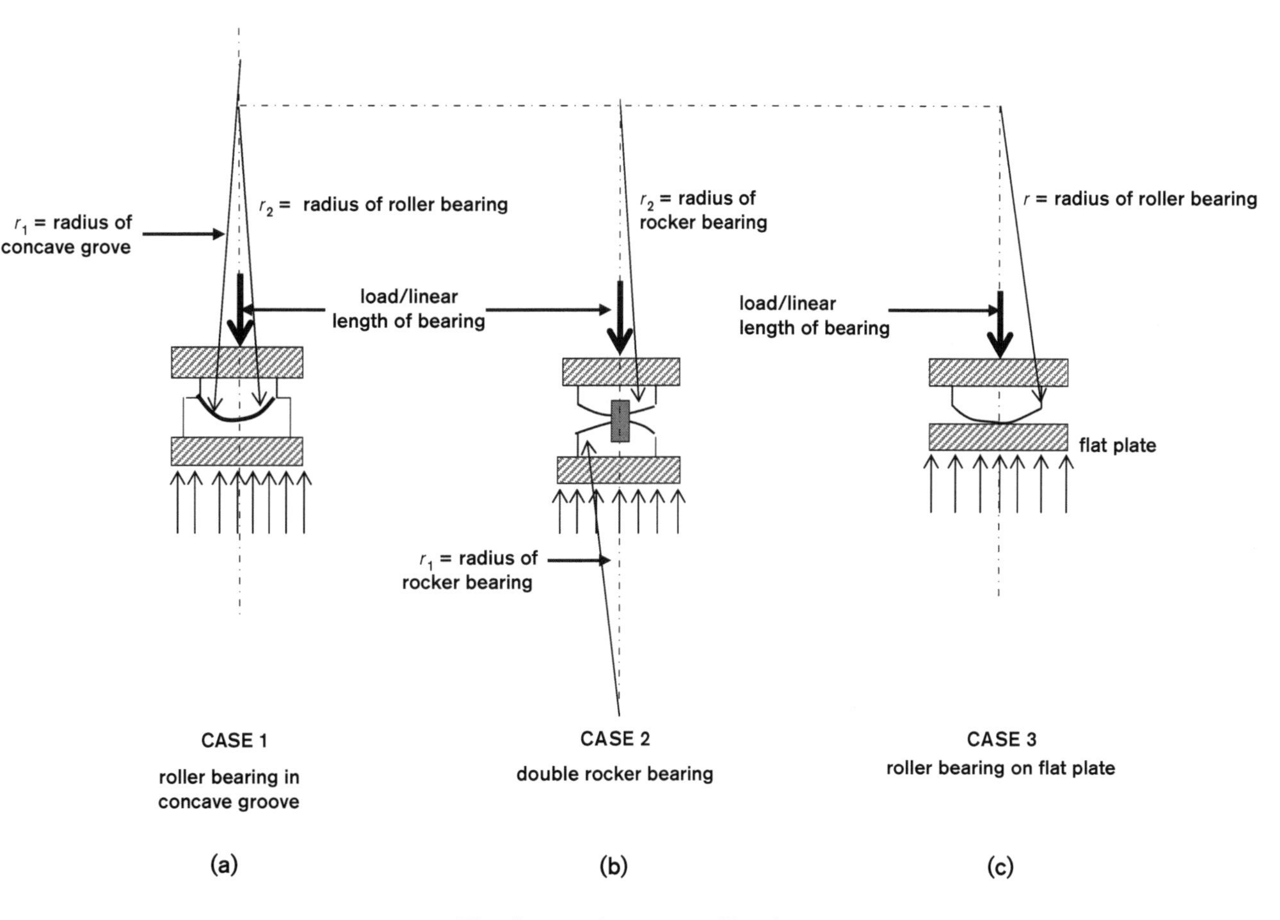

Fig. A.1. Various types of bearing

R_2 = radius of cylindrical roller,

The value of Poisson's ratio ν is assumed to be 0.3.

Case 2: for a cylindrical roller of radius R_2 on a rocker of radius R_1, the maximum contact stress is given by

$$\sigma_u = 0.418[pE(R_1 + R_2)/(R_2R_1)]^{0.5}.$$

Case 3: for a cylindrical roller of radius R on a flat plate, the contact stress is given by

$$\sigma_u = 0.418[pE/R]^{0.5}.$$

In our calculations, we consider case 3: a cylindrical roller of radius R on a flat plate (see Fig. A.2). To limit the deformation of the roller and the plate to an acceptable level, the contact stress is limited to 1.75 times the ultimate tensile strength f_y. Thus,

$$1.75\sigma_u = 0.418[pE/R]^{0.5}.$$

Hence

$$p = 17.53\sigma_u^2R/E.$$

In BS 5400, Part 9.1 (British Standards Institution, 2006), the factor 17.53 is rounded up to 18.00. For a roller of diameter D on a flat surface, the above equation is equivalent to

$$p = 8.76\sigma_u^2D/E.$$

As rounded up and quoted in BS 5400, Part 9.1,

$$\sigma_u^2 = pE/(18R),$$

where $R = D/2$ is the radius of a rocker on a flat surface. We assume that the length of the bearing is 950/2 − 100 = 375 mm, say 400 mm. The unfactored loads are 4062 kN for the crane girder live load reaction and 279 kN for the crane dead load (from the gantry girder calculations in Chapter 3). Using partial safety factors $\gamma_{Gj} = 1.35$ for the dead load and $\gamma_{Qk} = 1.5$ for the live load,

$$\text{total ultimate load on bearing} = 1.5 \times 4062 + 135 \times 279 = 6470 \text{ kN}.$$

Therefore

$$\text{load/mm length of bearing} = p(\text{act}) = 6470 \times 10^3/400 = 16\ 174 \text{ N}.$$

We assume that the radius of the rocker bearing R is 1200 mm. The modulus of elasticity E is equal to 210 000 N/mm^2. Referring to Table 3.1 of Eurocode 3, Part 1-1, for steel grade S 460, we find that f_y = 440 N/mm^2 for a thickness of 60 mm < 80 mm, and

$$\text{maximum contact stress} = \sigma_u = (pE/(18R))^{0.5}$$
$$= [16\ 174 \times 210\ 000/(18 \times 1200)]^{0.5} = 397 \text{ N/mm}^2 < f_y \text{ (440 N/mm}^2).$$

Therefore we adopt a rocker bearing of radius R = 1200 mm (see Fig. A.2).

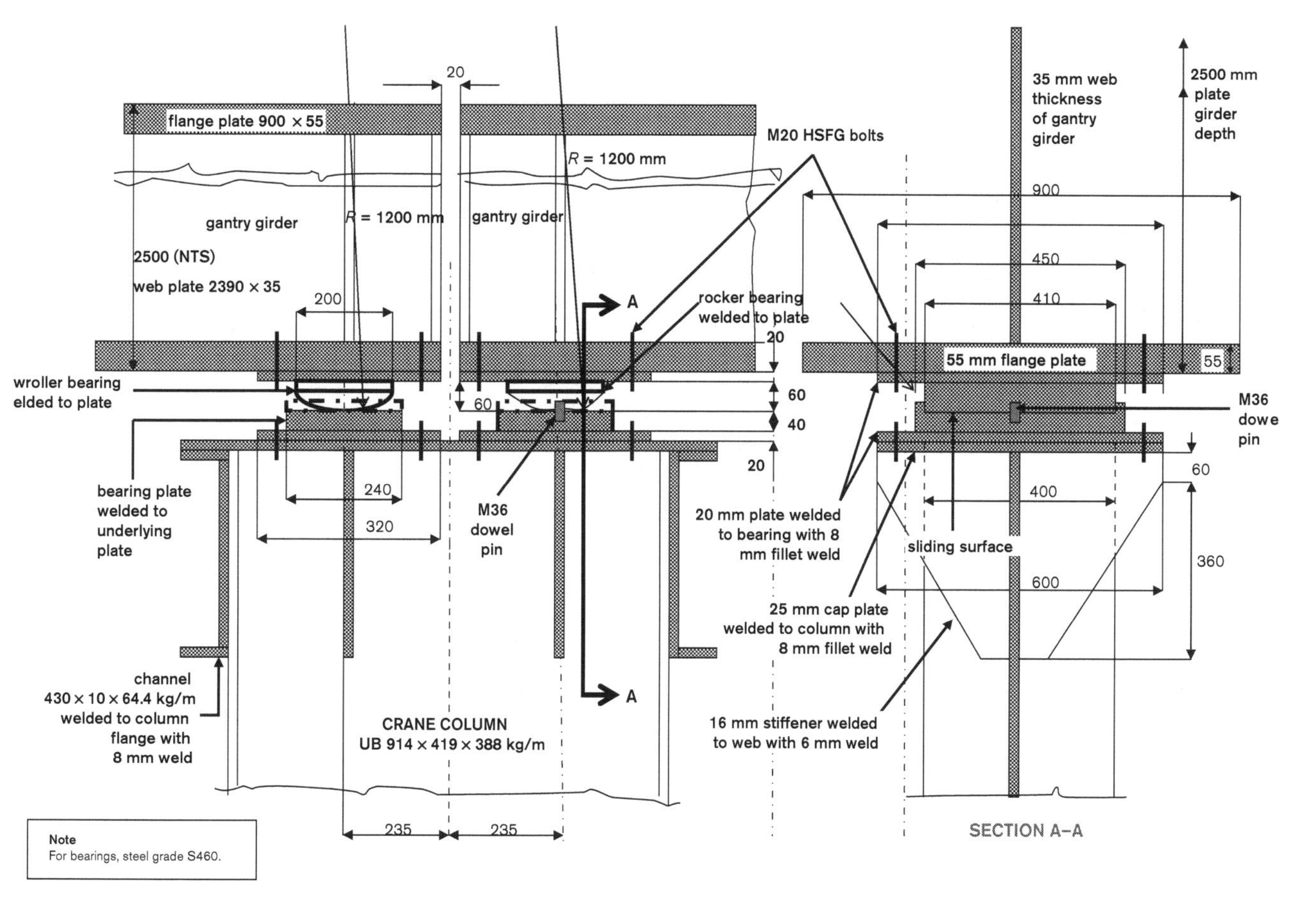

Fig. A.2. Details of bearing of gantry girder

References

British Standards Institution, 2006. BS 5400-9.1: 2006, Code of practice for design of bridge bearings.

Grinter, L.E., 1961. *Design of Modern Steel Structures*, Macmillan, New York.

Lee, D.J., 1990. *Bridge Bearings and Expansion Joints*, 2nd edn, Taylor & Francis, London.

APPENDIX B

Annex A of Eurocode 3, Part 1-1, BS EN 1993-1-1: 2005

In this appendix, we reproduce Tables 3.1, 5.2, 6.3 and 6.4 contained in Annex A of Eurocode 3, Part 1-1, and also Tables A1.1 and A1.2(B) of BS EN 1990: 2002(E).

Table 3.1. Nominal values of yield strength f_y and ultimate tensile strength f_u for hot rolled structural steel [from Eurocode 3, Part 1-1, BS EN 1993-1-1: 2005]

Standard and steel grade	Nominal thickness of the element t [mm]			
	$t \leq 40$ mm		40 mm $< t \leq 80$ mm	
	f_y [N/mm²]	f_u [N/mm²]	f_y [N/mm²]	f_u [N/mm²]
EN 10025-2				
S 235	235	360	215	360
S 275	275	430	255	410
S 355	355	510	335	470
S 450	440	550	410	550
EN 10025-3				
S 275 N/NL	275	390	255	370
S 355 N/NL	355	490	335	470
S 420 N/NL	420	520	390	520
S 460 N/NL	460	540	430	540
EN 10025-4				
S 275 M/ML	275	370	255	360
S 355 M/ML	355	470	335	450
S 420 M/ML	420	520	390	520
S 460 M/ML	460	540	430	530
EN 10025-5				
S 235 W	235	360	215	340
S 355 W	355	510	335	490
EN 10025-6				
S 460 Q/QL/QL1	460	570	440	550

Table 5.2 (sheet 1 of 3). Maximum width-to-thickness ratios for compression parts [from Eurocode 3, Part 1-1, BS EN 1993-1-1: 2005]

Internal compression parts

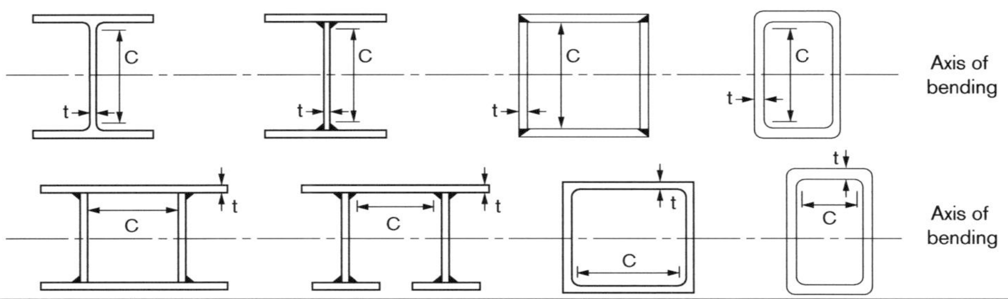

Axis of bending

Axis of bending

Class	Part subject to bending	Part subject to compression	Part subject to bending and compression
Stress distribution in parts (compression positive)	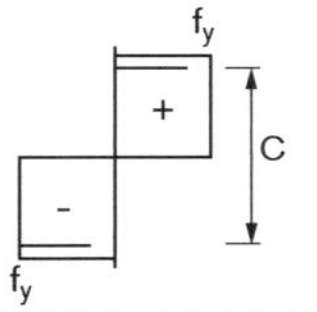	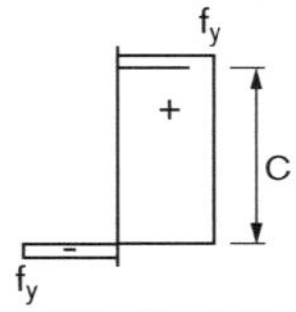	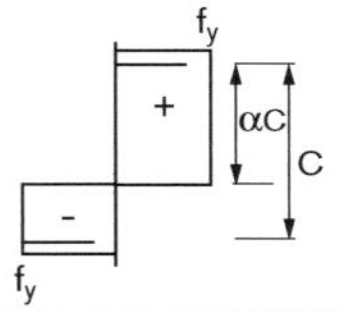
1	$c/t \le 72\varepsilon$	$c/t \le 33\varepsilon$	when $\alpha > 0.5 : c/t \le \dfrac{396\varepsilon}{13\alpha - 1}$ when $\alpha \le 0.5 : c/t \le \dfrac{36\varepsilon}{\alpha}$
2	$c/t \le 83\varepsilon$	$c/t \le 38\varepsilon$	when $\alpha > 0.5 : c/t \le \dfrac{456\varepsilon}{13\alpha - 1}$ when $\alpha \le 0.5 : c/t \le \dfrac{41.5\varepsilon}{\alpha}$
Stress distribution in parts (compression positive)	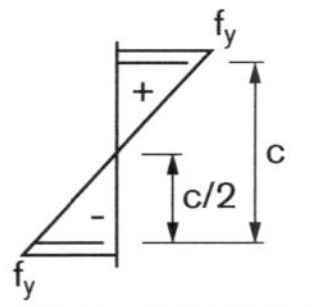	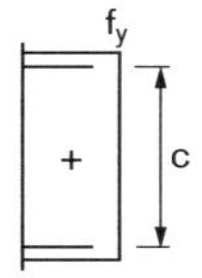	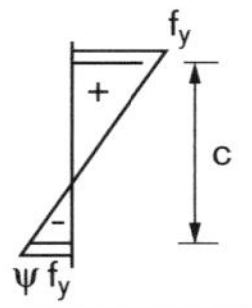
3	$c/t \le 124\varepsilon$	$c/t \le 42\varepsilon$	when $\psi > -1 : c/t \le \dfrac{42\varepsilon}{0{,}67 + 0{,}33\psi}$ when $\psi \le -1^{*)} : c/t \le 62\varepsilon(1-\psi)\sqrt{(-\psi)}$

$\varepsilon = \sqrt{235/f_y}$	f_y	235	275	355	420	460
	ε	1.00	0.92	0.81	0.75	0.71

* $\psi \le -1$ applies where either the compression stress $\alpha \le f_y$ or the tensile strain $\varepsilon_y > f_y/E$

Table 5.2 (sheet 2 of 3). Maximum width-to-thickness ratios for compression parts [from Eurocode 3, Part 1-1, BS EN 1993-1-1: 2005]

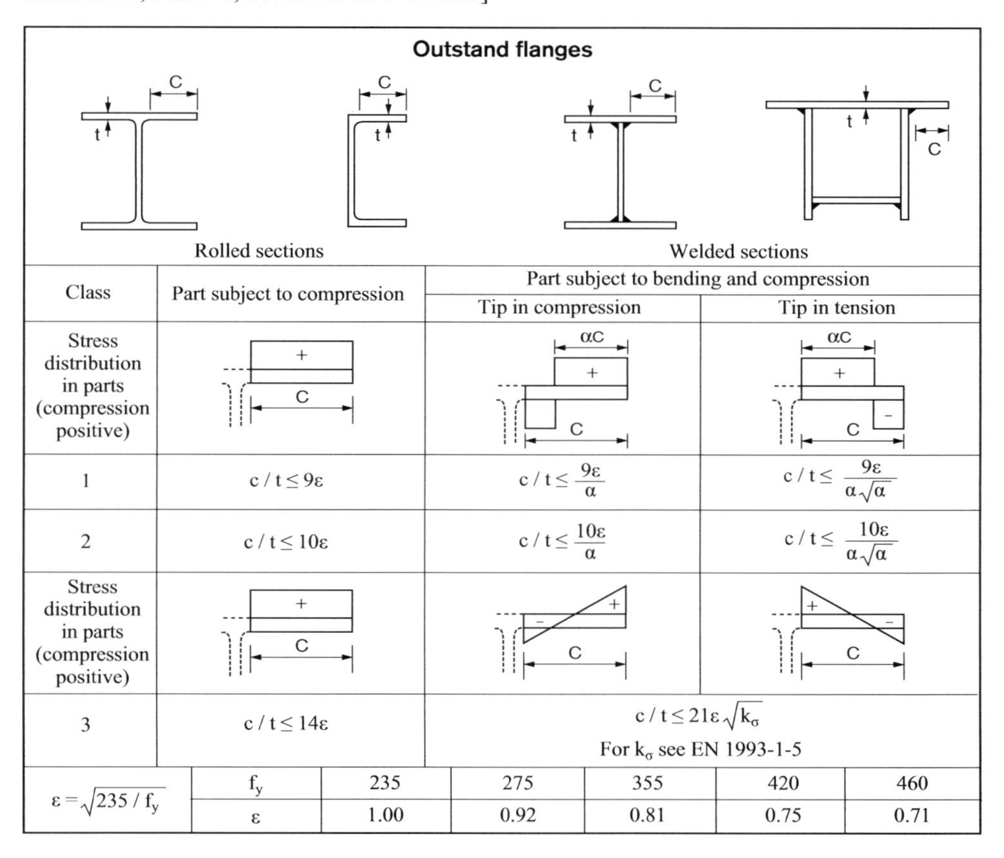

Outstand flanges

Class	Part subject to compression	Part subject to bending and compression: Tip in compression	Part subject to bending and compression: Tip in tension
Stress distribution in parts (compression positive)	+ c	αc + c	αc + − c
1	$c/t \le 9\varepsilon$	$c/t \le \frac{9\varepsilon}{\alpha}$	$c/t \le \frac{9\varepsilon}{\alpha\sqrt{\alpha}}$
2	$c/t \le 10\varepsilon$	$c/t \le \frac{10\varepsilon}{\alpha}$	$c/t \le \frac{10\varepsilon}{\alpha\sqrt{\alpha}}$
Stress distribution in parts (compression positive)	+ c	+ − c	+ − c
3	$c/t \le 14\varepsilon$	$c/t \le 21\varepsilon\sqrt{k_\sigma}$ For k_σ see EN 1993-1-5	

$\varepsilon = \sqrt{235/f_y}$	f_y	235	275	355	420	460
	ε	1.00	0.92	0.81	0.75	0.71

Table 5.2 (sheet 3 of 3). Maximum width-to thickness ratios for compression parts [from Eurocode 3, Part 1-1, BS EN 1993-1-1: 2005]

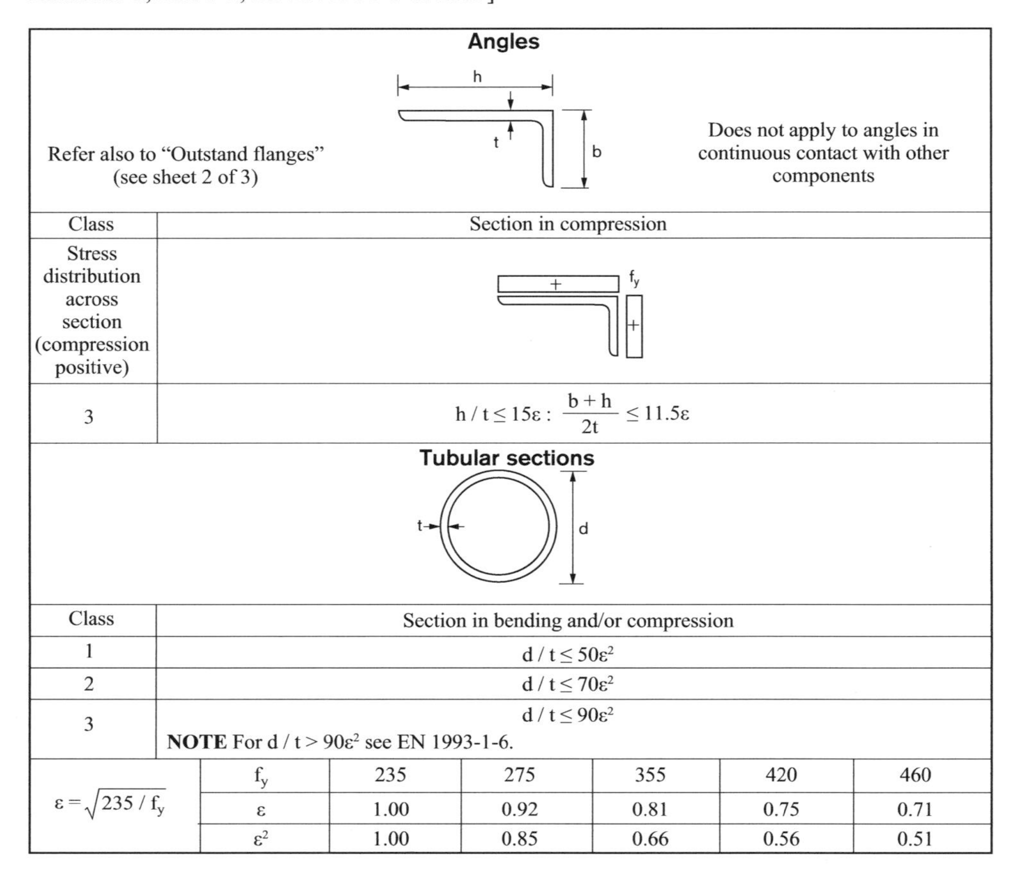

Angles

Refer also to "Outstand flanges" (see sheet 2 of 3)

Does not apply to angles in continuous contact with other components

Class	Section in compression
Stress distribution across section (compression positive)	
3	$h / t \leq 15\varepsilon : \frac{b+h}{2t} \leq 11.5\varepsilon$

Tubular sections

Class	Section in bending and/or compression
1	$d / t \leq 50\varepsilon^2$
2	$d / t \leq 70\varepsilon^2$
3	$d / t \leq 90\varepsilon^2$ **NOTE** For $d / t > 90\varepsilon^2$ see EN 1993-1-6.

$\varepsilon = \sqrt{235 / f_y}$	f_y	235	275	355	420	460
	ε	1.00	0.92	0.81	0.75	0.71
	ε^2	1.00	0.85	0.66	0.56	0.51

Table 6.3. Recommended values for imperfection factors for buckling curves [from Eurocode 3, Part 1-1, BS EN 1993-1-1: 2005]

Buckling curve	a	b	c	d
Imperfection factor α_{LT}	0.21	0.34	0.49	0.76

Table 6.4. Recommended values for lateral torsional buckling of sections using equation (6.56) [from Eurocode 3, Part 1-1, BS EN 1993-1-1: 2005]

Cross-section	**Limits**	**Buckling curve**
Rolled I-sections	$h/b \leq 2$ $h/b > 2$	**a** **b**
Welded I-sections	$h/b \leq 2$ $h/b > 2$	**c** **d**
Other cross-sections	–	**d**

Table A1.1. Recommended values of ψ factors for buildings

Action	ψ_0	ψ_1	ψ_2
Imposed loads in buildings, category (see EN 1991-1)			
Category A: domestic, residential areas	0.7	0.5	0.3
Category B: office areas	0.7	0.5	0.3
Category C: congregation areas	0.7	0.7	0.6
Category D: shopping areas	0.7	0.7	0.6
Category E: storage areas	1.0	0.9	0.8
Category F: traffic area, vehicle weight ≤ 30kN	0.7	0.7	0.7
Category G: traffic area, 30kN < vehicle weight ≤ 160kN	0.7	0.5	0.3
Category H: roofs	0	0	0
Snow loads on buildings (see EN 1991-1-3)*			
Finland, Iceland, Norway, Sweden	0.70	0.50	0.20
Remainder of CEN Member States, for sites located at altitude H > 1000 m a.s.1.	0.70	0.50	0.20
Remainder of CEN Member States, for sites located at altitude H ≤ 1000 m a.s.1.	0.50	0.20	0
Wind loads on buildings (see EN 1991-1-4)	0.6	0.2	0
Temperature (non-fire) in buildings (see EN 1991-1-5)	0.6	0.5	0

NOTE: The ψ values may be set by the National annex.
* For countries not mentioned below, see relevant local conditions.

Table A1.2(B). Design values of actions (STR/GEO) (Set B) [from BS EN 1990: 2002(E)]

Persistent and transient design situations	Permanent actions		Leading variable action	Accompanying variable actions*	
	Unfavourable	Favourable		Main (if any)	Others
Equation (6.10)	$\gamma_{Gj,sup}G_{kj,sup}$	$\gamma_{Gj,inf}G_{kj,inf}$	$\gamma_{q.1}Q_{k,1}$	–	$\gamma_{Q,1}\psi_{0,1}Q_{k,i}$
Equation (6.10a)	$\gamma_{Gj,sup}G_{kj,sup}$	$\gamma_{Gj,inf}G_{kj,inf}$	–	$\gamma_{Q,1}\psi_{0,1}Q_{k,1}$	$\gamma_{Q,1}\psi_{0,1}Q_{k,i}$
Equation (6.10b)	$\zeta\gamma_{Gj,sup}G_{kj,sup}$	$\gamma_{Gj,inf}G_{kj,inf}$	$\gamma_{q,1}Q_{k,1}$	–	$\gamma_{Q,1}\psi_{0,1}Q_{k,i}$

* Variable actions are those considered in Table A1.1.

The following values for γ and ζ are recommended when equations (6.10), (6.10a) and (6.10b) are used:

$\gamma_{Gj,sup} = 1.35$;

$\gamma_{Gj,inf} = 1.0$;

$\gamma_{Q,1} = 1.5$ where unfavourable (0 where favourable);

$\gamma_{Q,i} = 1.5$ where unfavourable (0 where favourable);

$\zeta = 0.85$ (so that $\zeta\gamma_{Gj,sup} = 0.85 \times 1.35 = 1.15$).

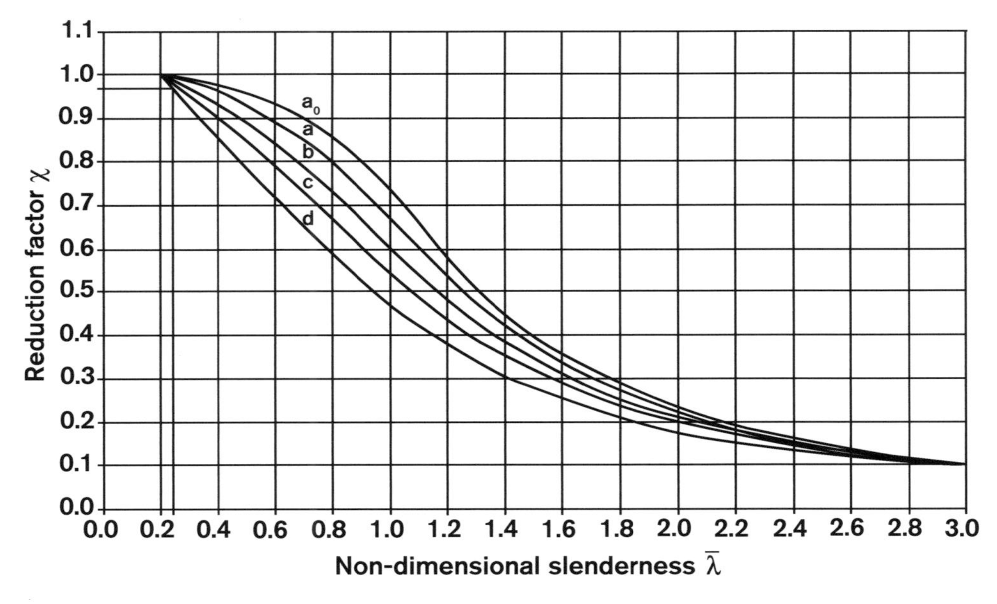

Buckling curves

Further Reading

Books

American Institute of Steel Construction, 2006. *Manual of Steel Construction,* 13th edn., American Institute of Steel Construction Inc, New York, USA.

Arya, C., 2009. *Design of Structural Elements: Concrete, Steelwork, Masonry and Timber Designs to British Standards and Eurocodes,* Spon Press, London, UK.

Butterworth, S., 1949. *Structural Analysis by Moment Distribution,* Longmans, Green & Co., New York, USA.

Fisher Cassie, W., 1951. *Structural Analysis: The Solution of Statically Indeterminate Structures,* Longmans, Green & Co., London, UK.

Gardner, L. and Nethercot, D.A., 2005. *Designer's Guide to EN 1993-1-1 Eurocode 3: Design of Steel Structures: General Rules and Rules for Buildings,* Thomas Telford, London, UK.

Gaylord, E.H., Gaylord, C.N. and Stallmeyer J.E., 1996. *Structural Engineering Handbook,* 4th edn., McGraw-Hill, New York, USA.

Ghosh, K.M., 2009. *Foundation Design in Practice,* Prentice-Hall India, New Delhi, India.

Grinter, L.E., 1961. *Design of Modern Steel Structures,* Macmillan, New York, USA.

Husband, J. and Harby, W., 1947. *Structural Engineering,* Longmans, Green & Co., London, UK.

Kani, G., 1957. *Analysis of Multistory Frames,* translated from German 5th edn, Frederick Unger, New York, USA.

Pippard, A.J.S. and Baker, J.F., 1953. *The Analysis of Engineering Structures,* Arnold, London, UK.

Salmon, C.G., Johnson, J.E. and Malhas, F.A., 2009. *Steel Structures: Design and Behavior,* 5th edn., Prentice Hall, New York, USA.

Salmon E.H., 1945. *Materials and Structures,* Vol. 1, Longmans, Green & Co., London, UK.

Salmon E.H., 1948. *Materials and Structures,* Vol. 2, Longmans, Green & Co., London, UK.

Seward, D., 2009. *Understanding Structures: Analysis, Materials, Design,* 4th revised edn., Palgrave Macmillan, Basingstoke, UK.

Spofford, C.M., 1939. *The Theory of Struc*tures, McGraw-Hill, New York, USA.

Steel Construction Institute, 1990. *Steel Work Design Guide to BS 5950: Section Properties and Member Capacities,* Pt. 1, vol. 1, 5th edn., Berkshire, UK.

Steel Construction Institute, 2003. *Steel Designers' Manual,* 6th edn., Blackwell Science, Oxford, UK.

Stewart, D.S.,1947. *Practical Design of Simple Steel Structures,* Vol. 1, Constable, London, UK.

Stewart, D.S., 1953. *Practical Design of Simple Steel Structures,* Vol. 2, Constable, London, UK.

Timoshenko, S.P. and Gere, J.M., 1961. *Theory of Elastic Stability;* McGraw-Hill Kogakusha, Tokyo, Japan.

Way, A.G.J. and Salter, P.R., 2003. *Introduction to Steelwork Design to BS 5950-1: 2000,* Steel Construction Institute, Berkshire, UK.

Westbrook, R. and Walker, D. 1996. *Structural Engineering Design in Practice,* 3rd edn., Longman, London, UK.

Wiegel, R.L., 1970. *Earthquake Engineering,* Prentice-Hall, New Jersey, USA.

Papers

Berrett, N., 2007. Structural design benefits sustainability, *The Structural Engineer,* 85(9).

Byfield, M.P. and Nethercot, D.A., 1997. Material and geometric properties of structural steel for use in design, *The Structural Engineer,* 75(21).

Cham, S.L., 2005. Codified simulation-based design of steel structures in the new Hong Kong steel code, *The Structural Engineer,* 83(20).

Dallard, P., Fitzpatrick, A.J., Flint, A., Le Bourva, S., Low, A., Ridsdill Smith, R.M. and Willford, M., 2001. The London Millennium Footbridge, *The Structural Engineer,* 79(2).

Davies, J.M., 2000. Steel framed house construction, *The Structural Engineer,* 78(6).

Eyre, J. and Croll, J., 2007. Serviceability design of columns, *The Structural Engineer,* 85(3).

Gardner, L. and Nethercot, D.A., 2004. Structural stainless steel design: a new approach, *The Structural Engineer,* 82(21).

Graham, J., 2003. An elastic–plastic (second order) plane frame analysis method for design engineers, *The Structural Engineer,* 81(10).

Hicks, S., 2007. Strength and ductility of headed stud connectors welded in modern profiled steel sheeting, *The Structural Engineer,* 85(10).

Jeffers, E., 1997. Design of braced steel I-beam assemblies without rigid decking, *The Structural Engineer,* 75(15).

Jenkins, W.M., 2001. A neural-network based reanalysis for integration with structural design, *The Structural Engineer,* 79(13).

Ji, T., 2003. Concepts for designing stiffer structures, *The Structural Engineer,* 81(21).

Lam, D., 2007. Strengthening of metallic structures using externally bonded FRP composite, *The Structural Engineer,* 85(6).

Lim, J.B.P. and Nethercot, D.A. 2002. Design and development of a cold-formed steel portal framing system, *The Structural Engineer,* 80(21).

Pang, P.T.C., 2006. Fire engineering design and post assessment, *The Structural Engineer,* 84(20).

Tsang, N., 2007. Research in structural engineering at Imperial College London, *The Structural Engineer,* 85(1).

Index